Some Aspects of Human and Veterinary Nutrition

World Review of Nutrition and Dietetics

Vol. 28

Series Editor
Geoffrey H. Bourne, Atlanta, Ga.

S. Karger · Basel · München · Paris · London · New York · Sydney

Some Aspects of Human and Veterinary Nutrition

Volume Editor
Geoffrey H. Bourne, Atlanta, Ga.

21 figures and 37 tables, 1978

S. Karger · Basel · München · Paris · London · New York · Sydney

World Review of Nutrition and Dietetics

Vol. 16: XXXII + 511 p., 25 fig., 44 tab., 1973. ISBN 3–8055–1398–4
Vol. 17: XII + 318 p., 70 fig., 60 tab., 1973. ISBN 3–8055–1336–4
Vol. 18: XII + 327 p., 18 fig., 29 tab., 1973. ISBN 3–8055–1438–1
Vol. 19: XIV + 319 p., 20 fig., 76 tab., 1974. ISBN 3–8055–1589–8
Vol. 20: XII + 347 p., 16 fig., 51 tab., 1975. ISBN 3–8055–1841–2
Vol. 21: X + 327 p., 31 fig., 37 tab., 1975. ISBN 3–8055–2133–2
Vol. 22: XII + 346 p., 28 fig., 50 tab., 1975. ISBN 3–8055–2135–9
Vol. 23: XII + 315 p., 5 fig., 21 tab., 1975. ISBN 3–8055–2243–6
Vol. 24: X + 262 p., 23 fig., 50 tab., 1976. ISBN 3–8055–2344–0
Vol. 25: X + 296 p., 53 fig., 69 tab., 1976. ISBN 3–8055–2363–7
Vol. 26: Human and Veterinary Nutrition. XII + 274 p., 22 fig., 21 tab., 1977.
ISBN 3–8055–2392–0
Vol. 27: Some Aspects of Human Nutrition. X + 178 p., 17 fig., 81 tab., 1977.
ISBN 3–8055–2393–9

Cataloging in Publication
    Some aspects of human and veterinary nutrition / volume editor, Geoffrey H. Bourne.
    – Basel, New York: Karger, 1978.
    (World review of nutrition and dietetics; v. 28)
    1. Nutrition 2. Animal Nutrition I. Bourne, Geoffrey Howard, 1909– II. Title
    III. Series
    W1 W0898 v. 28 / QU 145 S691
    ISBN 3–8055–2672–5

© Copyright 1978 by S. Karger AG, 4011 Basel (Switzerland), Arnold-Böcklin-Strasse 25
Printed in Switzerland by Thür AG Offsetdruck, Pratteln
ISBN 3–8055–2672–5

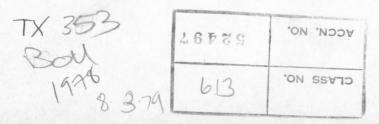

Advisory Board

Contents

Contents

Contents

Biochemistry and Physiology of Magnesium
Jerry K. Aikawa, Denver, Colo.

Contents

Bone Growth and Development in Protein-Calorie Malnutrition

John H. Himes, Yellow Springs, Ohio

Hepatocarcinogens in Nigerian Foodstuffs
Enitan A. Bababunmi, Anthony O. Uwaifo and Olumbe Bassir, Ibadan, Nigeria

Carcass Evaluation of Cattle, Sheep and Pigs
Alastair Cuthbertson, Milton Keynes

Contents

Wld Rev. Nutr. Diet., vol. 28, pp. 1–111 (Karger, Basel 1978)

Parenteral Nutrition

Alan Shenkin and Arvid Wretlind

Department of Human Nutrition, Medical Faculty, Karolinska Institutet, Stockholm

Contents

I. Introduction

During the last 40 years, parenteral or intravenous nutrition has become an important achievement in medicine. The developments in the field of parenteral nutrition are due to advances in three main areas. Firstly, our growing knowledge of mammalian physiology and biochemistry has resulted in a fairly comprehensive understanding of nutritional requirements in health and in disease states. Secondly, elaborate procedures have been evolved for the preparation of infusion solutions containing the various nutrients. And thirdly, there has been a considerable improvement in the surgical and nursing techniques used in the insertion and maintenance of intravenous catheters.

Even though the basic principles of parenteral nutrition are now firmly established, there are still many problem areas which await solution by future research. In the following review, we propose to discuss both those aspects of the field about which there is general agreement as well as a number of the areas of uncertainty. The literature on parenteral nutrition is now extensive and it is quite impossible to review all of it comprehensively in one article. The early literature has been well documented by *Geyer* (1960), and more recent accounts have been given in books by *Lee* (1974), *Ghadimi* (1975) and *Fischer* (1976a).

II. Historical Background

The first recorded intravenous infusion was as early as 1656, when Sir *Christopher Wren*, the architect of St. Paul's Cathedral in London, injected nutrients and drugs into the veins of dogs. He used a goose quill attached to a pig's bladder for the intravenous administration of an infusion solution containing ale, opium and wine. Soon afterwards, *Johann Major* (1662) and independently *Richard Lower* (*Lower and King,* 1662) made intravenous infusions and blood transfusions in animals, and *Jean Baptiste Denis* (1667) transfused blood from a lamb to a human being. Nothing was known at that time about microorganisms, sterility or immunological incompatibility, not to speak of osmolality or pyrogens. Consequently, it was not surprising that a large number of complications and deaths resulted from these experiments.

Technical improvements followed, and intravenous infusions of dextrose and saline were increasingly used. It was not until the end of the 19th century, however, after the investigations by Sir *Joseph Lister* (1870) on asepsis, and the discovery by *Louis Pasteur* (1877) of microbiological infection, that parenteral infusions could be performed with a reasonable chance of success. Further progress was slow until *Kausch,* in 1911, infused dextrose for nutrition after surgical procedures, and *Henriques and Anderson* (1913) injected enzymatically hydrolysed protein into animals. Fat was infused experimentally by *Murlin and Riche* in 1916. It was used in man as a result of *Yamakawa*'s work in 1920.

Elman (1937) and his co-workers demonstrated the effectiveness of parenteral infusion of casein hydrolysate for post-operative nitrogen (N) loss in surgical patients. It was soon realised that N equilibrium could only be obtained if an adequate amount of energy was given. Glucose or other water-soluble energy sources could not at that time be used to cover the energy need. To provide the necessary quantities of energy, comprehensive studies were undertaken by several teams of investigators in the late 1940s and early 1950s to produce a suitable fat emulsion for intravenous use. The fat emulsions of those early days, however, produced fever, chills, jaundice and coagulation defects, and were consequently withdrawn from clinical use in the United States in 1964. At the same time, safe and effective fat emulsions were being developed by *Wretlind* and co-workers in Sweden. After careful toxicological investigations, *Schuberth and Wretlind* in 1961 found that a fat emulsion containing soybean oil with egg yolk phospholipids as emulsifier was free from toxic action.

Parenteral nutrition with fat emulsion as the main energy source was rapidly developed for adults in several clinics in Scandinavia. Moreover, clinical use of fat emulsions in surgical paediatric patients was made by *Rickham* in 1967 and *Børresen and Knutrud* in 1969. This rapid and widespread acceptance of the technique was largely due to the practicality of giving the infusions into peripheral veins, a route of administration made possible by the vein sparing effect of fat emulsions.

Concurrently with the Swedish work, another method of parenteral nutrition was studied in the United States, where the new type of fat emulsion was not available. To provide high energy parenteral solutions, American investigators had to use highly concentrated glucose solutions. These hypertonic solutions damage the walls of the veins and cause phlebitis and thrombosis on administration in peripheral veins. To avoid these complications, *Dudrick et al.* (1969) succeeded in developing a system of central vein infusion to perform a total parenteral alimentation ('hyperalimentation') both in animal and man.

Once the major problems involved in safe infusion of adequate amounts of energy substrates and N had been overcome, the way was open for development of preparations of minerals and vitamins suitable for intravenous use (*Wretlind,* 1975). Thus, it is now possible to meet all the nutritional requirements of the individual by the intravenous route.

III. General Views on Intravenous Nutrition

A daily supply of energy and nutrients is necessary to maintain a patient in an optimal state of nutrition, and to offer the best resistance to illness and trauma, such as infections, burns, surgery, etc. If normal oral nutrition is difficult or impossible to maintain, essential nutrients may be provided, either by tube-feeding or by the intravenous route. Intravenous feeding should be resorted to, only when oral feeding and tube-feeding are impossible.

Intravenous nutrition may be regarded as an alternative to oral feeding. From this point of view, it is desirable to supply the nutrients intravenously, in the same quantity and form as they are normally transferred to the blood from the intestine after adequate oral feeding. The quantities of energy and nutrients supplied should be adjusted to cover the basal requirements, and to compensate for increased losses and previous deficiencies.

When food is digested and absorbed from the intestine, its water-soluble components are handled in different ways. Some are partly transformed and stored in the liver, whereas others are partly transformed without storage. Yet other substances pass untransformed into the general circulation. Fats circumvent the liver, being carried via the thoracic duct into the venous bloodstream. Thus, in intravenous nutrition, the fat emulsions are entering the general circulation in a similar way as the chylomicrons from natural fats. All other nutrients, however, reach the circulation in a more unnatural way, since they are discharged directly into the systemic rather than the portal circulation.

When all nutrients, absorbed from ordinary, adequate oral food, are given intravenously, this is termed *complete intravenous feeding.* According to *Dudrick et al.* (1968, 1969) the energy requirement may be adequately covered by glucose and amino acids with a concomitant supply of all other nutrients, but no fat. Because of the hypertonic solutions used, such an intravenous alimentation can only be performed via a central vein catheter. The complete intravenous nutrition including fat may be given either via a peripheral or a central vein catheter.

An intravenous nutrition, in which nutrients are supplied by the intravenous route alone, is called *total intravenous nutrition.* This term is therefore not synonymous with complete intravenous nutrition. In some cases, the oral intake is insufficient, indicating a requirement for *supplementary intravenous nutrition* with all, or some special nutrients.

There are several conditions where parenteral nutrition has been of decisive importance, and has often proved life-saving. Views on the application of intravenous nutrition have been summarised by *Peaston* (1968a) as follows: 'In recent years, materials have become available whereby intravenous nutrition can be adequately maintained. The therapeutic decision not to use them is a positive action to starve the patient, and must in itself be justified unless the complications from their use outweigh the advantages conferred.' In other words, this means that every patient should have the right to receive daily a sufficient supply of nutrients in one way or another. Total and supplementary intravenous nutrition should therefore be available in all hospitals handling seriously ill patients.

Dudrick and Rhoads (1971) even suggest that it may become possible to correct metabolic diseases, such as diabetes and atherosclerosis by parenteral nutrition. They also speculate on the possibility to treat patients with malignant disease using a diet which may starve the tumour but at the same time may support the patient's nutrition.

There is now a trend to use parenteral nutrition for exact metabolic studies. The supply of nutrients can be exactly determined, and there are no difficulties associated with variations in absorption, which always occur when nutrients are given enterally. In a number of investigations, the balances of N and several minerals have been determined more exactly than was previously possible. It is expected that such advanced investigations will permit more accurate quantitation of requirements for the various nutrients in different clinical conditions.

IV. Nutrients in Intravenous Nutrition

To maintain or restore the normal body composition of adults and to obtain normal growth in infants by complete intravenous nutrition, it is necessary to supply the nutrients or groups of nutrients (water, amino acids, fats, carbohydrates, minerals and vitamins) mentioned in table I.

The energy supplied should cover the resting metabolism, physical activity and a specific dynamic action of about 6% (*Swift and Fischer,* 1964). Depending upon their general condition, patients receiving parenteral nutrition can have very variable amounts of physical activity. This corresponds to an extra energy consumption of about 1.7 kJ/min (0.4 kcal/min) in a sitting position and 5.8 kJ/min (1.4 kcal/min) when walking (Recommended Dietary Allowances,

Table I. Nutrients essential for complete intravenous nutrition

Fluid	Water
Sources for synthesis of body proteins and energy	Amino acids Glucose (Fructose) Fat
Minerals	Sodium Potassium Calcium Magnesium Iron Zinc Manganese Copper Chromium Selenium Molybdenum Chloride Phosphorus Fluoride Iodide
Water-soluble vitamins	Thiamine Riboflavine Niacin Vitamin B_6 Folacin Vitamin B_{12} Pantothenic acid Biotin Ascorbic acid
Fat-soluble vitamins	Vitamin A Vitamin D Vitamin K_1 Tocopherol

1968). As summarised in table II, the energy requirements can therefore be expected to vary between 5.0 and 8.4 MJ (1,200 and 2,000 kcal/day, or between 91 and 130 kJ (22 and 31 kcal/kg body weight/day. In general, women require less energy than men (table II).

Table III shows the estimated energy requirements for intravenous nutrition during the period of growth. The energy supply to children on intravenous nutrition should be adjusted so as to obtain a normal weight increase.

Table II. Calculations of energy requirements of adults on intravenous nutrition

| | Energy requirements of *men* aged | | | | | | Energy requirements of *women* aged | | | | | |
| | 18–35 | | 35–75 | | over 75 years | | 18–55 | | 55–75 | | over 75 years | |
	kcal	MJ	kcal	MJ	kcal	MJ	kcal	MJ	kcal	MJ	kcal	MJ
Energy expenditure per 24 h Resting metabolism	1,600	6.7	1,500	6.3	1,350	5.6	1,300	5.4	1,200	5.0	1,100	4.6
Sitting in or beside bed 0–8h (8 × 60 × 0.4 = 192 kcal)	0–192	0–0.80	0–192	0–0.80	0–192	0–0.80	0–192	0–0.80	0–192	0–0.80	0–192	0–0.80
Walking 0–60 min (60 × 1.4 = 84 kcal)	0–84	0–0.35	0–84	0–0.35	0–84	0–0.35	0–84	0–0.35	0–84	0–0.35	0–84	0–0.35
Specific dynamic effect (6% of ingested energy)	102–120	0.43–0.50	95–113	0.40–0.47	86–104	0.36–0.44	83–99	0.35–0.41	77–94	0.32–0.39	70–88	0.29–0.37
Total energy expenditure	1,702–1,996	7.1–8.4	1,595–1,889	6.7–7.9	1,436–1,730	6.0–7.2	1,383–1,675	5.8–7.9	1,277–1,570	5.3–6.6	1,170–1,464	4.9–6.1
Body weight, kg	65		65		63		55		53		53	
Energy requirements/kg body weight/day	26–31	0.11–0.13	25–29	0.10–0.12	23–28	0.096–0.12	25–31	0.10–0.13	24–30	0.10–0.13	22–28	0.092–0.12

Table III. Recommended energy allowances for total intravenous nutrition to cover resting metabolism, growth, specific dynamic effect and some physical activity

Age, years	Energy supply[1]			
	kcal/kg	MJ/kg	kcal/day	MJ/day
0–1	120–90[2, 3]	0.50–0.38	500–1,000[2]	2.1–4.2
1–7	90–75[3]	0.38–0.31	1,000–1,500	4.2–6.3
7–12	75–60[3]	0.31–0.25	1,500–2,000	6.3–8.4
12–18	60–30[3]	0.25–0.13	2,000	8.4

[1] The values on left indicate the supply at the lowest age of the interval.
[2] *Dudrick et al.* (1969); *Bφrresen et al.* (1970a, b).
[3] *Reissigl* (1965).

A. Water

The body can tolerate a loss of fluid for only a short time. After 3–4 days without fluid, symptoms of dehydration usually appear. In adults, the loss of water is then 4–10% of the body weight. With a loss of 20–25% of body weight (40% of total body water) survival is not possible. Children are more sensitive to changes in body water. A loss of body water amounting to only 13–16% of body weight is not compatible with survival.

During the first week after birth, the daily milk consumption of the normal baby rises from 0 to 130 ml/kg, and during the following weeks to 150–180 ml/kg. A baby weighing 3 kg, thus consumes 400 ml water/day, which is about 20% of its total body water, as almost 70% of the new-born's body consists of water. The high water metabolism of the infant makes a relatively small quantitative disturbance of great importance.

Under normal conditions, an adequate daily amount of water is considered to be 100–150 ml/kg body weight for neonates and infants, 30–120 ml/kg for children, and 30 ml/kg for adults. Because of the fairly close relationship in most individuals between intake of energy substrates and water, it has also proved convenient for practical purposes to express the water requirement in terms of the energy intake. For infants, this is 0.4 ml/kJ or 1.5 ml/kcal, whereas for adults it is 0.25 ml/kJ or 1 ml/kcal (Recommended Dietary Allowances, 1968). Thus, the total daily intake of water should approximately amount to 0.5–2 litres for neonates, infants and children, and 1.5–2 litres for adults, in the absence of fever or other causes of abnormal losses of water.

A large number of factors influence the loss of fluid which makes it difficult to estimate the minimum need of water in disease states. However, a combination of careful clinical examination together with well kept fluid balance charts should allow reasonably accurate estimates to be made in most cases.

When calculating the fluid balance, the metabolic water formed by the combustion of protein, carbohydrates and fat must be taken into account. According to *Consolazio et al.* (1963), metabolism of 1 g protein, carbohydrate or fat gives rise to 0.41, 0.60 or 1.07 ml water, respectively. From 1 g of an amino acid mixture, 0.52 ml water is formed [(0.41 + 0.23)/1.23 = 0.52; 1 g protein equals 1.23 g amino acid mixture, 0.23 g of which is 'bound' water]. Thus, a supply of 8.4 MJ (2,000 kcal) representing 50 g amino acids, 140 g carbohydrates and 140 g fat corresponds to 260 ml metabolic water.

B. Protein and Amino Acids

1. Some Biochemical and Physiological Aspects on the Metabolism of Proteins and Amino Acids

Under normal conditions, there is a dynamic equilibrium between the synthesis of body proteins and the breakdown of proteins in the body. Some of the amino acids are deaminated, and then metabolised in a variety of ways depending upon the type of amino acid and the general metabolic conditions, the final products being carbohydrates, fat, carbon dioxide, water, urea, and energy. A small proportion may be excreted unchanged in the urine. When amino acids are administered intravenously, they enter directly the body's amino acid pool and are then metabolised in the normal physiological way (fig. 1).

Blood, erythrocytes, plasma, albumin and amino acid mixtures have been and are used for protein nutrition intravenously. In table IV the indications for these 5 products are summarised (*Bünte*, 1964).

It is not possible to use haemoglobin or globin for a proper protein nutrition. The reason for this is that this protein contains too little of the essential amino acid isoleucine. The supply of large quantities of blood also incurs the risk of overloading the vascular system, of infections, immunisation, haemosiderosis and inhibition of the bone marrow (*Jordal*, 1965). Consequently, blood should only be given to replace blood which has been lost.

Between 50 and 70% of infused albumin enters the interstitial space and the half-life of albumin is about 20 days (*Artz*, 1959; *Beeken et al.*, 1962). Consequently, albumin is a very inefficient source of amino acids for repletion of other tissue proteins and it should only be used when treating patients suffering from hypoalbuminaemia.

Only with *amino acid mixtures* is it possible to provide a *physiological intravenous protein alimentation*. In this way, if the amino acid composition is correct, the amino acid pool should receive the same supply of amino acids as through absorption of amino acids from the intestinal tract.

a) The role of the digestive tract. Food of vegetable or animal origin is eaten and degraded by the digestive enzymes. The initial hydrolysis products of the proteins (tripeptides, dipeptides and amino acids) are absorbed by the mucosal cells of the intestine in which they are further hydrolysed, yielding a mixture of

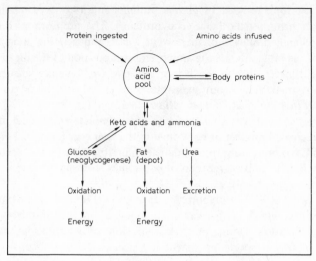

Fig. 1. A summary of the metabolism of amino acids and proteins.

Table IV. Indications for intravenous infusions of products containing protein and amino acids

Product	Indications
Blood	Blood losses
Erythrocytes	Normovolaemic anaemia
Plasma	Losses of plasma protein
Albumin	Hypoalbuminaemia
Amino acids	Protein nutrition

the component amino acids. Already in the mucosal cells a part of the glutamic and aspartic acids participate in a transamination reaction with pyruvic acid, producing ketoglutaric acid, oxaloacetic acid and alanine. The outflow of amino acids from the gut into the portal vein following protein ingestion is characterised by a predominance of alanine and the virtual absence of glutamate and aspartate (*Felig*, 1975). Nonetheless, most of the amino acids absorbed in the intestine are conveyed intact by the portal vein to the liver. There is, however, also an intensive synthesis of protein in the intestinal mucosa (*Lang*, 1972).

b) The role of the liver. In the liver, amino acids are metabolised in several ways. *Elwyn* (1970) found that in rats, after a meal rich in protein, 57% of the ingested amino acids were oxidised to urea, 23% passed into the general circulation, 6% were used for synthesis of plasma proteins and 14% were retained temporarily as liver protein ('the labile liver protein'). The proportion of the amino acid intake handled in each of these ways will vary depending upon the nutritional state of the organism and the amino acid composition of the diet.

The amino acid metabolism in the liver is very selective. Leucine, isoleucine and valine are not transformed, but pass into the general circulation to be metabolised mainly in muscles and kidney (*Miller*, 1962).

During the protein synthesis after ingestion of a complete amino acid mixture, there is an increased number of polysomes in the liver cells (*Munro*, 1972). If the amino acid mixture, however, is deficient in tryptophan, the protein synthesis is arrested, with a disaggregation of polysomes resulting in an accumulation of inactive monosomes in the cells (*Fleck et al.*, 1965).

Hallberg and Soda (1976) has reported that amino acids increase the blood flow in the splanchnic vessels. In this way, more blood is conveyed to the portal circulation during digestion. During parenteral infusions of amino acids, an increased portal blood flow supports the metabolic processes just described.

It is now clear that there is a considerable flux of amino acids between the various organs, and particularly between skeletal muscle and liver (*Felig*, 1975). Alanine is of special interest because it can be synthesised in muscle from pyruvate (from glucose metabolism) and amino groups (from branched-chain amino acid oxidation), and transferred to the liver where the N is converted to urea and the carbon skeleton is reconverted to glucose (glucose-alanine cycle).

c) Blood amino acid levels. The liver seems to serve as a buffer to protect other organs from the effect of excessive concentrations of amino acids. The regulating action of the liver controls the plasma amino acid concentration to 0.3–0.4 mmol/l (= 4.4 mg/100 ml). If the level falls, amino acids from the liver protein are mobilised to cover the need. The amino acid concentration in serum is influenced by hormones, such as insulin, glucagon, growth hormone and glucocorticoids. There seems to be a feedback system by which the secretion of the hormones is controlled by the amino acid content in the serum (*Munro*, 1971). Insulin release is stimulated by the branched-chain amino acids, and glucagon by the non-essential amino acids (*Cahill*, 1972a).

The plasma amino acid profile in normally nourished individuals has been well documented (*Scriver et al.*, 1971). Characteristic changes in this profile take place in protein malnutrition, in particular the proportion of branched-chain amino acids (especially valine) falling, whilst the concentration of the non-essential amino acids is maintained or may even be slightly elevated (*Waterlow and Harper*, 1975). Studies of plasma amino acid concentrations have been used to assess the dietary requirement of amino acids (*V.R. Young et al.*, 1971).

Abitbol et al. (1975) have described the plasma aminogram of low birth weight infants and compared it to that of the full-term infant and adult. In the low birth weight infants, hypoaminoacidaemia was present at birth. The concentrations of glutamine, alanine, glycine, histidine, and ornithine were significantly below the values in full-term infants. By contrast, tyrosinaemia was found in the low birth weight infants. Of some interest was the observation that the plasma tyrosine level declined to normal when ascorbic acid in an amount of 100 mg/ day was supplemented.

The plasma aminograms in patients after trauma were investigated by *Dölp et al.* (1975). Before any infusions, the plasma levels of amino acids were generally lower than normal. When an amino acid mixture was infused, all of the amino acid levels were increased. Methionine, isoleucine, leucine and glycine showed striking increases above the normal range. The cysteine level did not reach the normal value. Since the amino acid mixture used did not contain any cystine, the probable explanation of this finding is that of formation of cystine from methionine was impaired. The plasma amino acid profile also shows typical changes in patients with severe liver disease (sect. VIII.F).

It should, however, be pointed out that the concentrations of the various amino acids in plasma may not reflect the composition of the complete amino acid pool. Studies on muscle biopsy samples have demonstrated that the intra-cellular concentration of free amino acids is usually higher and has a very different profile from plasma (*Bergström et al.*, 1974).

2. Amino Acid Imbalance, Antagonism and Toxicity

Amino acids show many very complex interrelations, such as *amino acid imbalance, amino acid antagonism,* and *amino acid toxicity* (*Harper and Rogers,* 1965).

Amino acid imbalance has been studied in animals, especially in rats. Imbalance can occur either if an amino acid mixture has a reduced amount of an essential amino acid or if some amino acid is present in excess of a limiting amino acid. The imbalance can usually be compensated by the addition of the limiting amino acid (*Harper and Rogers,* 1965). *Harper* (1970) found that amino acid imbalance causes loss of appetite. This loss of appetite can be reproduced by intravenous injection of the imbalanced mixture. *Robergs and Leung* (1973) showed that the depressed food intake was associated with a change in plasma level of individual amino acids.

Amino acid antagonism is different from imbalance (*Harper and Rogers,* 1965). It is compensated by addition of an amino acid which is not limiting, but is chemically related to the amino acid present in excess.

Large amounts of single amino acids added to diets induce various *toxic reactions* including depression of growth. Large amounts of tyrosine lead to damage in the paws and eyes of rats. Methionine causes pathological changes in several organs. The only kind of amino acid intoxication seen in parenteral nutrition has been ammonia intoxication resulting from the use of preparations containing large amounts of glycine (*Doolan et al.*, 1965). *Harper et al.* (1956)

had previously shown that the infusion of glycine in dogs produced a marked rise in the blood ammonia levels with no corresponding increase in urea. A protective effect of casein hydrolysate against the toxicity of glycine had been demonstrated by *Handler et al.* (1949). *Najarian and Harper* (1956) found that, when arginine was given concomitantly with glycine, there was no significant rise in the blood ammonia levels. An appreciable increase in urea was recorded, indicating that the urea was probably formed from the ammonia.

3. Requirements of Amino Acids

a) Requirement of total amino acid N. The requirements of amino acid N can be calculated from known N losses. The daily losses for an individual with basal energy requirement have been summarised in Recommended Intakes of Nutrients for the United Kingdom (1969) as follows:

Endogenous N loss in urine	= 2 mg/basal kcal
Minimum faecal N loss	= 0.57 mg/basal kcal
Minimum sweat, skin and hair N loss	= 0.08 mg/basal kcal
Total N loss	2.65 mg/basal kcal

The losses thus amount to a total of 2.65 mg N/basal kcal, equivalent to 17 mg (2.65 × 6.25) of a protein with a biological value of 100, or 20 mg (2.65 × 7.5) of an adequate amino acid mixture/basal kcal. This means that 56–66 mg N (2.65 × 21 to 2.65 × 25 kcal) of an adequate amino acid mixture/kg/day should cover the minimum requirement of protein in adults. The coefficient of variation of the required N quantity for balance is probably between 10 (*Garrow,* 1969) and 15% (*Hussein et al.,* 1968) of the N quantity consumed. A supply of 73–86 mg N (0.55–0.65 g amino acids)/kg, which is 30% above the average minimum requirements, would thus cover the basal requirement of protein in almost all adults (Recommended Dietary Allowances, 1968). After compensating a loss of amino acids through the urine of up to 10%, one can expect to require 80–95 mg N (0.60–0.71 g amino acids)/kg/day to cover the basal needs. The quantity calculated in this way equals that which was sufficient to obtain slightly positive nitrogen balance in a patient extensively investigated by *K. Bergström et al.* (1972) during a 7-month period of complete intravenous nutrition.

According to the results and calculations given, a daily amino acid allowance of at least 95 mg N, or 0.7 g amino acids/kg, may be recommended for adults with basal energy requirements. This corresponds to 6–6.5 g N, or 50 g amino acids, for persons weighing 70 kg. This quantity is similar to the N supply from protein (calculated as protein with an NPU value of 100) present in a British or

Swedish average national diet with a relatively low energy supply of 6.3–8.4 MJ/day (1,500–2,000 kcal/day).

The basal requirement of oral protein in neonates and infants up to the age of 12 months has been calculated at 1.2–2.2 g/kg/day depending upon the age of the infant (*Fomon*, 1967). This corresponds to 1.5–2.7 g amino acid mixture.

Børresen and Knutrud (1969) and *Dudrick et al.* (1969) have intravenously given to post-operative neonates and infants 0.40–0.53 g N, or 3–4 g amino acids/kg body weight/day. With this amino acid supply, satisfactory growth and a positive N balance were obtained.

b) Requirements of individual essential and non-essential amino acids. The protein-synthesising system in all cells needs a complete mixture of 20 amino acids to produce the various proteins of the body. Eight of the amino acids (table V), so-called essential amino acids, cannot be synthesised by the adult body. Consequently, they have to be supplied in food, or in some other way.

For infants, histidine has also been found to be an essential amino acid. *Kofranyi et al.* (1969) have indicated that histidine prevents liver derangement in men. Investigations by *Bergström et al.* (1970) have shown that histidine also seems to be necessary for an optimal utilisation of amino acid mixtures in patients with uraemia. The capacity of the body to synthesise arginine is limited. Thus, arginine must be included in amino acid mixtures for intravenous nutrition in order to obtain optimal utilisation of the other amino acids supplied.

In the fetus and immature baby, there is a lack of cystathionase activity in the liver which prevents the conversion of methionine to cystine and cysteine (*Sturman et al.*, 1970). Cystine is therefore an essential amino acid for these cases. *Stegink and den Besten* (1972) administered amino acid solutions free from cysteine to 8 healthy men intravenously or by nasogastric tubes. When the solutions were given intravenously, the plasma cystine concentration dropped markedly. When feeding was switched to the enteral route, the concentration rose, but returned to baseline only when a cystine-containing diet was fed. These studies indicate that the synthesis of cysteine from methionine is limited, even in the adult subject.

The phenylalanine hydroxylase system is not fully developed in prematures. This means that phenylalanine cannot be converted into tyrosine. An intravenous supply of an amino acid mixture including phenylalanine, but without tyrosine, results in a decrease of the tyrosine concentration in serum. Thus, amino acid solutions for intravenous nutrition of prematures and infants should contain tyrosine as well as phenylalanine (*Panteliadis et al.*, 1975).

Jürgens and Dolif (1972) have shown that the utilisation of intravenous amino acid preparations is higher if alanine, proline and glutamic acid are included in the amino acid mixture. According to *Panteliadis et al.* (1975), glycine is necessary in small amounts (0.2 g/kg) for neonates on parenteral nutrition to maintain normal serum levels of glycine and serine.

Table V. Nutritional value of amino acids for intravenous alimentation

Isoleucine Leucine Lysine Methionine Phenylalanine Threonine Tryptophan Valine	Essential in all conditions[1]
Arginine	Necessary for optimal utilisation of amino-acid mixtures and for detoxification[2]
Cysteine-cystine	Essential for the fetus, and necessary for the maintenance of normal plasma level of cystine in adults[3]
Glycine	Necessary for neonates[4]
Histidine	Essential for infants and in uraemia[5] Necessary to prevent liver derangement[6]
Tyrosine	Essential for neonates[4,7]
Alanine Glutamic acid Proline	Necessary for optimal utilisation of amino acid mixtures[8]
Aspartic acid Serine	Source of non-specific N
Glutamine Asparagine	Produced *in vivo* by amidation of glutamic and aspartic acids; no requirement in intravenous nutrition

[1] *Rose* (1957).
[2] *Najarian and Harper* (1956); *Malvy et al.* (1961).
[3] *Sturman et al.* (1970); *Stegink and den Besten* (1972).
[4] *Panteliadis et al.* (1975).
[5] *Holt and Snyderman* (1961); *Bergström et al.* (1970).
[6] *Kofranyi et al.* (1969).
[7] *Jürgens and Dolif* (1972).
[8] *Jürgens and Dolif* (1968); *Dolif and Jürgens* (1970).

Hitherto, there have been no investigations of the specific effects of aspartic acid and serine on the utilisation of an intravenous amino acid mixture. The nutritional value of the various amino acids is summarised in table V.

The absence of adequate amounts of the non-essential amino acids in the food can be partly compensated by providing a suitable source of N, which will enable the body to synthesise these amino acids. Although N-containing substances, such as diammonium citrate, urea, or non-essential amino acids such as glycine or alanine can act as such N sources, these substances do not lead to an equal and adequate synthesis of all the various non-essential amino acids. It is clear that the most effective way of ensuring that optimal amounts of all non-essential amino acids are present in the cells is to supply such a mixture of the non-essential amino acids (*Swenseid et al.*, 1959).

A mixture for intravenous protein nutrition therefore requires the two components: the essential amino acids and the non-essential amino acids. The proportion of the essential amino acids to the total N is of great nutritional importance, and may be expressed as grams of the essential amino acids per gram of total N (E/T ratio) (FAO/WHO Expert Group, 1965).

The requirements for the individual essential amino acids have been investigated in various connections. Previously, research workers have usually adopted the recommendations stated by *Rose* in 1957. Later investigations have demonstrated pronounced differences in the requirements for infants and adults. A summary of these studies is shown in table VI. *Steffee et al.* (1950) investigated the repletion of a protein-depleted rat with different amino acid mixtures. An effective repletion could only be obtained with an amino acid mixture containing 2–5 times the quantities of essential amino acids required for maintenance. If a similar situation exists in protein-depleted man where intravenous nutrition is required, using the data from table 4 in the paper by *Steffee et al.* (1950) and the figures given by *Munro* (1972) of the requirement of amino acids to maintain the body proteins, an estimate can be made of amino acid requirement for repletion in man (table VI).

In patients where intravenous nutrition is indicated, it is usually desirable that body proteins are repleted. Thus, an amino acid mixture is required with the same high content of essential amino acids as occurs in body proteins or other proteins with a high biological value (about 45–50% essential amino acids corresponding to an E/T ratio of about 3). Expressed differently, the amino acid mixture optimal for the growing infant seems to be the most appropriate for repletion of protein in the adult.

4. Amino Acid Preparations

Based on the present knowledge, several amino acid solutions for parenteral infusion have been composed. Table VII summarises some of the most important factors to be considered in relation to the composition of intravenous amino acid preparations.

The two general types of amino acid preparations for intravenous alimentation are *protein hydrolysates* and *mixtures of crystalline amino acids.*

Table VI. The requirement of essential amino acids in human subjects of various ages. All the investigations mentioned have been made with oral feeding

Requirement	Infants	Child 10–12 years	Adult women	Adult men		Estimated for repletion
	(Holt et al.)[1] mg/kg	(Nakagawa et al.)[1] mg/kg	(Hegsted)[1] mg/kg	(Rose)[1] mg/kg	(Inoue et al.)[1] mg/kg	(Steffee et al.)[2] mg/kg
Histidine	25	–	–	–	–	7.4
Isoleucine	111	28	10	10	11	10
Leucine	153	49	13	11	14	39
Lysine	96	59	10	9	12	54
Methionine and cystine	50	27	13	14	11	16
Phenylalanine and tyrosine	90	27	13	14	14	57
Threonine	66	34	7	6	6	11
Tryptophan	19	3.7	3.1	3.2	2.6	4.3
Valine	95	33	11	14	14	20
Total essential amino acids	705	261	80	81	85	219

[1] Results summarised by *Munro* (1972).
[2] The values have been calculated from table 4 in the paper by *Steffee et al.* (1950). The maintenance for man according to *Inoue et al.* (*Munro,* 1972) has been used as basis for the estimation of repletion requirement of amino acids in man. The amount of histidine is given in relation to the calculated amount of tryptophan.

Table VII. Considerations on intravenous amino acid mixtures

Quantities of essential amino acids
Quantities of non-essential amino acids
Ratio of essential amino acids to total N
Percentage essential amino acids of total amino acids
Ratio of an individual essential amino acid to total N
Ratio of an individual essential amino acid to total essential amino acids
D- and *L*-isomers of the amino acids; only *L*-forms should be accepted
Imbalance and antagonism; proof that the mixture is well-balanced
Toxicity: glycine content
Biological value in experimental tests on animals
Biological value in man
Urinary loss

Table VIIIa. Amino acid content of egg protein and some commercial amino acid preparations. The amino acid values are given in grams/16 g N (T) of the protein or of the amino acid mixtures

L-Amino acids	Quantity of amino acid in 16 g N				
	egg protein	aminosol[1]	aminofusin L-Reihe[2]	amino-norm[3]	amino-plasmal L5[3]
Isoleucine	6.6	6.5	3.3	2.6	5.1
Leucine	8.8	11.4	4.6	4.5	8.9
Lysine	6.4	10.1	4.2	7.2	5.6
Aromatic amino acids	10.0	7.5	4.6	4.7	6.4
Phenylalanine	5.8	6.4	4.6	4.2	5.1
Tyrosine	4.2	1.1	–	0.3	1.3
Sulphur-containing amino acids	5.5	5.5	4.4	4.5	4.3
Methionine	3.1	3.7	4.4	4.4	3.8
Cysteine-cystine	2.4	1.8	–	0.1	0.5
Threonine	5.1	5.0	2.1	4.1	4.1
Tryptophan	1.6	1.3	1.0	2.4	1.8
Valine	7.3	8.7	3.2	5.9	4.8
E = Total amount of essential amino acids/16 g N	51.3	56.0	27.4	35.8	41.0
Alanine	7.4	4.1	12.6	9.4	13.7
Arginine	6.1	4.3	8.4	5.0	9.2
Asparagine	–	–	–	2.6	3.3
Aspartic acid	9.0	8.6	–	0.8	1.3
Glutamic acid	16.0	27.6	19.0	4.3	4.6
Glutamine	–	–	–	19.6	–
Glycine	3.6	2.4	21.1	4.9	7.9
Histidine	2.4	3.2	2.1	4.4	5.2
Ornithine	–	–	–	2.4	3.2
Proline	8.1	13.4	14.7	5.2	8.9
Serine	8.5	5.6	–	2.0	2.4
E/T ratio	3.2	3.5	1.7	2.24	2.6
E, % of total amino acids	46	45	26	37	41

[1] Vitrum, Stockholm. [2] J. Pfrimmer, Erlangen. [3] B. Braun, Melsungen.

Tables VIIIa and b show the composition of some commercial amino acid preparations for intravenous nutrition, as well as the aminogram of egg protein.

    a) Protein hydrolysates. Woodyatt et al. in 1915 and *Rose* in 1934 suggested the parenteral use of amino acids, but it was not until 1937 that *Elman,* having

Table VIIIb. Amino acid content of some commercial amino acid preparations. The amino acid values are given in grams/16 g N (T) of the protein or of the amino acid mixtures

L-Amino acids	Quantity of amino acid in 16 g N				
	FreAmine[1]	Intramin forte[2]	Trophysan[3]	Sohamin[4]	Vamin[5]
Isoleucine	7.6	4.6	1.6	8.1	6.6
Leucine	9.9	7.2	2.6	12.2	9.0
Lysine	7.9	5.2	5.8	18.7	6.6
Aromatic amino acids	6.1	7.2	2.2	11.7	10.2
Phenylalanine	6.0	7.2	2.2	–	9.4
Tyrosine	–	–	–	–	0.8
Sulphur-containing amino acids	6.0	7.2	2.6	8.3	5.6
Methionine	5.8	7.2	2.6	–	3.2
Cysteine-cystine	0.2	–	–	–	2.4
Threonine	4.4	3.3	1.6	8.6	5.1
Tryptophan	1.7	1.7	1.0	3.7	1.7
Valine	7.2	5.2	2.2	7.8	7.3
E = Total amount of essential amino acids per 16 g N	50.8	41.6	19.6	79.1	52.1
Alanine	7.7	15.0	–	–	5.1
Arginine	4.0	9.4	1.8	11.0	5.6
Asparagine	–	–	–	–	–
Aspartic acid	–	8.1	–	–	7.0
Glutamic acid	–	–	–	–	15.3
Glutamine	–	–	–	–	–
Glycine	23.0	9.4	59.8	7.3	3.6
Histidine	3.1	5.6	–	4.2	4.1
Ornithine	–	–	–	–	–
Proline	12.2	9.4	–	–	13.8
Serine	6.4	4.7	–	–	12.8
E/T ratio	3.2	2.6	1.23	4.9	3.2
E, % of total amino acids	47	40		78	44

[1] McGaw Lab., Glendale, Calif. [2] Astra, Södertälje. [3] Egic, Montargis. [4] Tanabe, Seijaku Co., Osaka. [5] Vitrum, Stockholm.

investigated the use of casein hydrolysate, initiated the modern use of intravenous amino acid alimentation.

One enzymatic casein hydrolysate which has been widely used in Europe is *Aminosol Vitrum*. The preparation is produced by a special method including dialysis, by which all the high molecular weight peptides and other substances which might cause allergic reactions are removed. It contains about 67% free amino acids and 33% low-molecular weight peptides (*Wretlind,* 1947). *Lidström and Wretlind* (1952) have shown that this dialysed, enzymatic casein hydrolysate has, in man, a biological value of about 90. When protein hydrolysate is infused, the amino acid N level in the blood is increased, and is rapidly reduced when the infusion is completed. *Lidström and Wretlind* (1952) found that 4.3% of the free, and 21% of the peptide-bound α-amino N supplied, were excreted via the urine in man.

Another protein hydrolysate, *Aminosol Abbott,* prepared by acid hydrolysis of fibrin, has also been shown to be well utilised. The studies by *Dudrick et al.* (1969) and *Kaplan et al.* (1969) have shown that casein and fibrin hydrolysate are approximately equal in effectiveness, as measured by rate of growth in infants on total intravenous nutrition.

Hyperammonaemia was found by *Johnson et al.* (1972) in 7 infants who were receiving casein or fibrin hydrolysate intravenously. In these patients, the authors suggest that the high concentration of ammonium ion in the protein hydrolysates may have contributed to the hyperammonaemia. This seems to indicate that amino acid mixtures should be free from ammonium ion.

The enzymatic protein hydrolysates have now been used for more than 30 years. A large number of investigations have demonstrated the clinical value of the use of these hydrolysates (*Tweedle et al.,* 1972).

b) Crystalline amino acid preparations. Mixtures of crystalline amino acids offer the advantages of flexibility in composition. Their relatively high cost, however, has limited their practical use. In 1940, *Shohl and Blackfan* first used a complete mixture of crystalline amino acids for parenteral nutrition. N balance in infants was promoted equally well by this mixture, or by a commercial protein hydrolysate. Since then, intravenous alimentation with mixtures of crystalline amino acids has been extensively studied.

Mixtures of amino acids with a high content of glycine have been used by some investigators. In general, few side reactions have been reported. *Malvy et al.* (1961) have, however, observed that a commercial amino acid preparation, with *DL*-amino acid and a high content of glycine, produced an ammonia coma in man. This is what might be expected because of the tendency of glycine to produce ammonia as earlier described. *Heird et al.* (1972c) have also observed hyperammonaemia in 4 infants receiving total parenteral nutrition, including the crystalline amino acid preparation FreAmine (table VIIIb), which has a relatively high content of glycine. The hyperammonaemia and the associated clinical signs were corrected by the administration of arginine and/or ornithine.

It has been suggested that it could be advantageous to compose amino acid solutions according to plasma amino acid pattern, instead of following the results of balance experiments in normal males. This principle has been followed by *Knauff* (1970), who claims to have obtained superior results in both normal persons and patients with duodenal ulcers.

Crystalline amino acid mixtures are well utilised for the synthesis of body protein when the requirement of essential amino acids is covered and the amino acid mixtures are well balanced. The best results in man have been obtained when the preparations were virtually complete, containing all essential and 10–12 non-essential amino acids (*Jacobson and Wretlind*, 1970).

5. Effect of Various Amino Acid Preparations on N Balance and Body Weight

a) In animals. The effect of infusion of amino acid preparations has been thoroughly investigated in dogs. *Dudrick et al.* (1967) using an acid fibrin hydrolysate, and *Håkansson et al.* (1967) using an enzymatic casein hydrolysate, demonstrated that these preparations are satisfactory N sources as part of total parenteral nutrition, leading to positive N balance and normal growth in Beagle puppies. In a similar way, *Holm et al.* (1975) have compared the utilisation of some crystalline amino acid preparations. The degree of utilisation varied depending on the concentration of the essential amino acids in the various solutions, as well as the ratio of essential amino acids to total N in the mixture.

Using complete intravenous nutrition in rats, *Wretlind and Roos* (1975) have observed marked differences in the efficiency of utilisation of different commercially available crystalline amino acid preparations.

b) In adult patients. Many studies testify to the ability of protein hydrolysates of either casein (*Schärli*, 1965; *Peaston*, 1967; *Jeejeebhoy et al.*, 1976) or fibrin (*Drudrick et al.*, 1969; *Long et al.*, 1976) to maintain or produce positive N balance and gain in body weight in adults, provided an adequate supply of energy substrates is simultaneously provided. This has also been shown to be possible with many of the available synthetic amino acid preparations (*K. Bergström et al.*, 1972; *Gazzaniga et al.*, 1975).

Controlled comparisons of crystalline amino acid solutions and enzymatic protein hydrolysates have been performed in a few cases. These studies have shown that different commercially available crystalline preparations can have markedly different effects on N balance. *Peaston* (1968b) showed that mixtures of racemic *DL*-amino acids were not as effective as a casein hydrolysate in improving N balance. Moreover, *Tweedle et al.* (1972) found that a casein hydrolysate was as effective as the best crystalline preparation in their study (Vamin). On the other hand, *C.L. Long et al.* (1974) showed that a fibrin hydrolysate was not as well utilised as the crystalline preparation FreAmine, probably because of a limiting content of valine and phenylalanine.

c) In infants. By using acid protein hydrolysate as N source, *Dudrick et al.* (1969), performed total intravenous nutrition by central venous catheter for

7–400 days in a large number of neonates after surgical treatment of serious congenital anomalies of the gastrointestinal tract. Satisfactory growth and positive N balance were obtained. *Heird and Winters* (1975) have used FreAmine in similar groups of infants as part of central vein infusions and also have obtained satisfactory growth, although they concluded that this preparation was somewhat deficient in arginine and was prone to produce acidosis.

Complete intravenous nutrition through peripheral veins in neonates and infants has also been studied in detail by *Børresen and Knutrud* (1969) and *Grotte* (1971). The amino acid solutions used were Aminofusin and Vamin (tables VIIIa, b). With these crystalline amino acid preparations also, growth as well as positive N balance were maintained. Details of these investigations are given in section IX.

d) Conclusions. The choice of amino acid preparations for intravenous use is wide. The current trend as towards the increased use of crystalline amino acid solutions. Although their composition is more accurately defined, in many cases probably little difference in overall N retention is obtained by the use of a well-balanced crystalline preparation as opposed to a casein or fibrin hydrolysate. Because of the considerable difference in cost of these preparations, the precise clinical situations where one type of preparation is superior, both in terms of N utilisation as well as freedom from side-effects, should be clearly defined in future studies. One factor which may influence the choice of the solution is the electrolyte content. Protein hydrolysates generally have a much higher sodium content than the crystalline preparations. This may be advantageous in patients with high sodium requirements, but equally it may be a drawback in certain patients, and hence it must be carefully considered when deciding the N source to be used.

6. Comparison of Enteral and Intravenous Supply of Amino Acids

The alimentary tract and the liver may protect the organism from excessive amounts of amino acids in food. Consequently, it is conceivable that amino acid solutions for parenteral use should have a different composition from the aminogram of the dietary proteins. There is, however, now very extensive clinical experience demonstrating a good utilisation of parenteral amino acid solutions with a wide variety of chemical composition. The most likely explanation of the evident success of parenteral nutrition seems to be that the infusions are given slowly so that the organs are able to metabolise the amino acids and other nutrients in approximately the normal way. Because the liver receives a substantial part of the total blood circulation, all the infused amino acids soon reach the liver, and appropriate modulations can take place.

A direct comparison between enteral and intravenous nutrition in rats has been made by *Levin et al.* (1975). Groups of rats with intragastric or central vein catheters were used. The infusions of various amino acid solutions were administered for 20 h/day during a 10-day period. The animals had free access to a N-free diet. There were no significant differences in the voluntary feed intake, weight gain or urinary N excretion between the rats given the amino acids intravenously or intragastrically. Administrations of unbalanced amino acid mixtures reduced weight gain and increased the urinary N excretion to the same extent in both groups of rats.

Peaston (1967) in a series of studies, performed complete parenteral nutrition in adult patients. The average daily supply was 100 g amino acids, as an enzymatic casein hydrolysate (Aminosol), 225 g carbohydrates, as glucose and fructose, and a total of 200 g fat (Intralipid) amounting to 3,200 kcal. The same positive N balance was obtained during both intravenous and oral feeding, with the same nutrients and amount of energy. *Johnston and Spivey* (1970) also reported similar results. Thus, an enzymatic protein hydrolysate seems to be utilised to the same extent whether given intravenously or enterally.

7. Some Factors Influencing Utilisation of Amino Acid Mixtures

a) Temporal relationships between infusions of amino acids and energy substrates. For an optimal utilisation of an amino acid mixture, it is of great importance that all essential amino acids are given simultaneously, whether the administration is intravenous or oral (*Geiger,* 1951). In view of what is known about protein synthesis in the cells, it is readily understandable that the ribosomes cannot continue to build up the protein chain when one or more amino acids is not present at the appropriate time. Under these circumstances, the amino acids would be deaminated and oxidised, especially in the liver.

Since protein synthesis requires the expenditure of considerable amounts of energy for efficient incorporation of amino acids into protein, a source of energy should be supplied at the same time as the amino acids. When fat emulsions are given with amino acids and other nutrients to patients on total intravenous nutrition, the daily infusions are often given either in day-time or night-time during 10–14 h (*Jacobson and Wretlind,* 1970). Amino acids, in combination with glucose as the only source of energy, must always be given very slowly; the daily infusions generally require a period of 24 h (*Dudrick et al.,* 1972a, b) because of the high osmotic pressure of the amino acid-glucose solutions used. In a recent study on 3 patients receiving long-term total parenteral nutrition, *Broviac et al.* (1976) showed that there was a small but significant decrease in the efficiency of use of amino acids when the infusion time was shortened from 24 to 12 h, but no further effect on N balance was observed when the infusion period was decreased to 8 h.

b) Relative amounts of energy substrates and amino acids. One question about which there is still no general agreement is the optimal ratio of energy to N for most effective utilisation of an amino acid mixture. In most cases, it probably lies between 150 kcal/g N, as recommended by *Moore* (1959) and

200 kcal/g N as used by *Johnston et al.* (1972). However, present evidence suggests that the greater the degree of hypercatabolism, the lower should be the energy:N ratio, possibly falling to as low as 120 kcal/g N in cases of very severe hypercatabolism (*Kinney,* 1976).

Bozzetti (1976) has studied the N balance in patients both during the post-operative period and following simple fasting. He found that it was necessary to give a larger amount of energy per gram amino acids in the post-operative period than during simple starvation to obtain the same utilisation of the amino acids supplied. These results support other observations that in patients suffering from malnutrition, i.e. with small stores of protein, it is possible, with a moderate supply of amino acid N, to obtain a positive N balance even with a small energy supply. Thus, the worse the patient's nutritional condition, the better the utilisation of the dietary protein or amino acids. Apparently this is because in individuals in a poor nutritional state, protein metabolism shifts towards anabolism. Under such conditions, an exogenous supply of energy, particularly in the form of carbohydrate, has also a more marked effect in reducing protein breakdown than in a healthy person.

c) Type of energy substrate. One particularly controversial area has been how the energy should be supplied to ensure maximum possible protein synthesis. Three main types of regime have been suggested: in one of these, no exogenous energy is provided and the organism uses its own stores of energy substrate, primarily fat (*Blackburn et al.,* 1973a, b). In the other regimes, exogenous energy is supplied, either in the form of carbohydrate alone (*Dudrick et al.,* 1969) or as carbohydrate plus fat (*Wretlind,* 1972).

Blackburn et al. (1973a, b) have suggested the intravenous infusion of amino acids without any glucose or other energy source in periods in which a negative N and energy balance is anticipated. A less negative N balance was produced compared with intravenous nutrition with glucose alone. The energy requirement was covered by mobilisation of body fat and the ketone bodies produced by catabolism of fatty acids. *Hoover et al.* (1975) have confirmed these observations over the 6-day post-operative period, but showed that the utilisation of the infused amino acids was only about 47%. In one non-surgical patient, the utilisation of the infused amino acids was found to be 72%. Presumably, the amino acids are less efficiently used in stressed patients. Although *Blackburn et al.* (1973a, b) initially postulated that the effectiveness of such amino acid infusions may result from a decreased stimulation of insulin secretion in comparison with glucose infusions, recent studies by *Greenberg et al.* (1976) have cast some doubt on this. In these latter studies, the addition of hypocaloric amounts of glucose or lipid to the amino acid infusion had no effect on N balance, despite producing significantly different responses in plasma insulin. It seems likely therefore that the effect of the 'amino acid only' regimes on N balance is a function of the amino acids themselves (*Felig,* 1976). Despite the intense interest in their mechanism of action, the clinical situations in which amino acid only regimes are indicated seem restricted to those cases in which it is desired to wait for just a few days before commencing complete parenteral nutrition.

With regard to the use of carbohydrate alone, or carbohydrate plus fat as energy source, it is clear that carbohydrate in particular has a very important effect on amino acid metabolism. Thus, the early oral studies of *Munro and Thompson* (1953) demonstrated the increased N retention produced by increasing the carbohydrate content of the diet, this being associated with a fall in the amino acid concentration of the plasma. It is probably that this effect is largely due to increased insulin secretion since insulin is essential for the uptake of amino acids, particularly into skeletal muscle, for synthesis of body proteins. *Fitzpatrick et al.* (1975) showed that continuous parenteral glucose (600– 740 g/day) led to a marked reduction in urinary N excretion, the lowest levels being reached after 4 days. This type of glucose infusion is associated with a marked rise in plasma insulin concentration (*Sanderson and Deitel,* 1974). Furthermore, *Hinton et al.* (1971) have demonstrated the protein sparing effect of high doses of intravenous glucose (600 g/day) plus exogenous insulin in patients with burns.

Apart from this effect upon insulin secretion, carbohydrates have a further effect on protein metabolism not shared by fats, since carbohydrates but not fat can yield the intermediates in metabolism which otherwise have to be obtained by breakdown of tissue proteins and catabolism of the constituent amino acids.

It is clear therefore that carbohydrate alone is a more effective source of energy for use in protein metabolism than fat alone. However, when a diet contains an isocaloric substitution of only a proportion of the carbohydrate with fat, in most cases little effect is observed on N balance. *Munro* (1951, 1964) has extensively reviewed this evidence with respect to oral studies. More recently, a number of investigators using intravenous infusions have also come to a similar conclusion. For example, *Jeejeebhoy et al.* (1976) have shown that only over the first 4 days of the infusion did a regime in which all non-protein energy was supplied by carbohydrate lead to a better N balance than an isocaloric regime in which 83% of the non-protein energy was provided by the soybean oil emulsion Intralipid. After that time, the regimes were equivalent in their effect on N balance. Several other studies have shown that regimes in which all of the non-protein energy was supplied by carbohydrate and regimes containing variable amounts of lipid did not show significant differences in their effects on N balance (*Gazzaniga et al.,* 1975; *Bark et al.,* 1976; *Broviac et al.,* 1976).

However, *J.M. Long et al.* (1974) found that in short-term studies in patients following trauma, the protein-sparing effect was related only to the amount of carbohydrate in the infusion with no effect on N balance associated with added lipid. It is possible that this may have been due to an increased adaptation to carbohydrate as energy substrate in a situation of particularly high energy requirement.

In summary, a parenteral nutrition regime containing carbohydrate as the sole source of non-protein energy leads to a greater effect on N utilisation than an isocaloric regime containing carbohydrate plus lipid when the infusion is only given for a few days. This effect may be particularly noticeable in very hyper-catabolic patients. With longer infusion periods, the two types of regime are virtually equivalent in their effects on N balance.

8. Summary of Measures to Ensure Optimal Effect of Intravenous Nutrition with Amino Acids

The amino acid preparation used should contain a well-balanced mixture of all the essential amino acids as well as most or all the non-essential amino acids.

For an optimal utilisation of the amino acids given intravenously, it is impera-tive also to meet the energy requirement by infusion. The best utilisation of an amino acid mixture seems to occur when the ratio between the supply of energy and the amino acid N is 120:1 to 200:1 kcal/g. Amino acids and non-protein energy should be given simultaneously to obtain an optimal nutritional effect. At the same time, it is important to ensure an adequate supply of all other nutrients.

The amount of N spared is generally better per energy unit of carbohydrate than per energy unit of fat. High levels of carbohydrate may be especially effective in promoting N retention in hypercatabolic patients, particularly if exogenous insulin is added to the infusion. In the patient who has only moderately elevated energy requirements, adaptation to lipid in an infusion which has continued for several days leads to virtually equivalent N utilisation with varying proportions of lipid, up to a maximum of 80% of the non-protein energy being in the form of lipid.

C. Carbohydrates, Polyols and Ethanol
1. Introduction

A minimum of 100 g carbohydrate/day, or 20 kcal% is necessary to avoid ketosis and increased protein catabolism (*Gamble*, 1958; *C.M. Young et al.*, 1971). This carbohydrate is used not only as a source of energy for certain tissues, but also to provide an adequate level of metabolic intermediates required for many biochemical reaction sequences. The carbohydrates which have been studied in relation to parenteral nutrition are glucose, fructose and maltose. The polyols, sorbitol, xylitol and glycerol which can be fairly readily converted to 'true' carbohydrates have also been investigated. Ethanol has been proposed as a possible 'carbohydrate' energy source, but it cannot be converted to glucose nor can it fulfil the other carbohydrate function of providing intermediary metabo-lites. A summary of the metabolic interrelationships of these various substances is shown in figure 2.

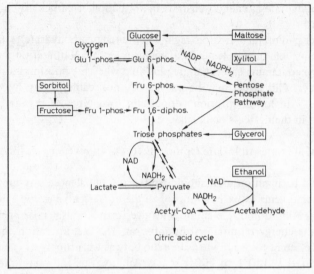

Fig. 2. The metabolic interrelationships of the various carbohydrates used in parenteral nutrition.

2. Carbohydrates

a) Glucose. Every cell in the human body has the capacity to oxidise glucose. In the liver, which is the main tissue in which other carbohydrates can also be metabolised, these substances are 'plugged' into biochemical pathways which were primarily evolved to handle glucose metabolism. In most tissues, metabolism of glucose proceeds largely via the Embden-Meyerhof (glycolytic) pathway leading to pyruvate and, under anaerobic conditions, to lactate. These substances can then be further oxidised in the citric acid cycle, yielding large amounts of energy in the form of ATP. In the liver and adrenal cortex, a variable proportion of the glucose can also be metabolised via the pentose phosphate pathway, which is mainly responsible for maintaining adequate supplies of NADPH, a coenzyme essential for many synthetic reactions.

In the normally nourished individual, glucose is of particular importance as the energy source for brain metabolism, since no other carbohydrate can substitute for this function of glucose. The brain requires about 120–140 g glucose/day, and any shortage in dietary supply of glucose will be supplemented by gluconeogenesis from amino acids. However, prolonged intake of an inadequate level of glucose will lead to the progressive adaptation of brain to use the ketone bodies, especially β-hydroxybutyric acid as an energy source (*Cahill and Aoki*, 1972). Nonetheless, since red blood cells have an obligatory requirement for 30–40 g glucose/day for anaerobic glycolysis, and since nervous tissue will always require at least about 20–30% of its energy in the form of glucose, if an adequate glucose intake is not maintained, increased gluconeogenesis from amino acids must ensue.

Insulin is essential for uptake of glucose into skeletal muscle cells. Intravenous infusion of glucose in amounts up to 0.5 g glucose/min in adults is associated with an insulin response adequate to prevent hyperglycemia and glycosuria (*Sanderson and Deitel,* 1974). However, in the post-operation or post-trauma period, a condition of insulin resistance may develop, necessitating the addition of exogenous insulin to the regime in some individuals (*Allison,* 1974). In spite of the glucose intolerance, there is little evidence that glucose turnover or oxidation is decreased after injury except very early in the acute phase (*Stoner,* 1970). Indeed, *Kinney et al.* (1970a, b) have shown that glucose oxidation may actually be increased after severe trauma.

Dudrick et al. (1968, 1969, 1972b), have a vast experience of post-operative parenteral nutrition with glucose as the only non-protein, energy source. In one of their publications (*Dudrick et al.,* 1969), the treatment of 300 severe, surgical adult cases, such as gastrectomy, colectomy, etc., were reported. The results showed rapid healing, positive N balance and weight increase. The infused nutrients were 525–750 g glucose (about 10 g/kg body weight) and 100–130 g amino acids supplemented with electrolytes and vitamins.

The amounts of glucose were increased in paediatric patients to cover the higher energy demand per kg (*Wilmore et al.,* 1969). Up to 25 g glucose/kg/day were given to infants operated for bowel atresias and ruptured omphaloceles. In spite of the severe operations and the large amounts of glucose, the results were good and no complications from glucose intolerance were reported. *Das et al.* (1970) studied an infant operated for an omphalocele. 27 g glucose/kg/day was given and blood glucose levels ranged between 2.5 and 6.0 mmol/l (44–108 mg/ 100 ml). Less than 1% of the total glucose infused was lost in the urine.

When fat emulsions are used to provide a proportion of the energy, simultaneous or subsequent glucose infusions are well metabolised in both adults (*Lee,* 1974) and infants (*Grotte et al.,* 1976). The value of glucose as carbohydrate in parenteral nutrition has therefore been amply proven.

b) Fructose. Because of the glucose intolerance in trauma and shock, it has been thought to be of value to find a substitute for glucose that could be metabolised independently of insulin. To a certain extent fructose does this.

Fructose does not cause hyperglycaemia and, therefore, does not invoke an insulin response (*Froesch and Keller,* 1972). It is more rapidly transformed into glycogen in the liver cells. Moreover, fructose also inhibits gluconeogenesis in the liver and, consequently, spares amino acids in glucose intolerance (*Froesch and Keller,* 1972).

Fructose enters the fat cells independently of glucose and insulin (*Froesch and Keller,* 1972). In these cells, large quantities of ketohexokinase oxidize the fructose and, thus, have an antiketogenic effect (*Bässler*, 1970a; *Froesch and Ginsberg*, 1962; *Thorén*, 1963b).

On the other hand, fructose as such cannot be used by the brain. About 70% of the fructose is transformed into glucose (*Ashby et al.*, 1965). The part of fructose which is transformed into glucose requires for its further metabolism the presence of insulin (*Keller and Froesch*, 1972).

The use of fructose as energy source may, however, be associated with a number of side-effects. The very rapid metabolism of fructose may lead to an accumulation of intermediary products such as lactic and pyruvic acid (*Mendeloff and Weichselbaum*, 1953; *Thomas et al.*, 1972). These and other results indicate that rapid and large infusions of fructose are contraindicated in patients with a tendency towards acidosis. Fructose, in amounts of less than 0.5 g/kg/h, seems to cause only small, or insignificant, changes in the lactate concentration in the blood. Also, as a result of the rapid and uncontrolled phosphorylation of fructose, fructose infusions may lead to a fall in ATP, ADP and inorganic phosphate content of the liver, with accumulation of AMP which can be further metabolised, leading to loss of adenosine nucleotides and the development of hyperuricaemia (*Fox and Kelley*, 1972).

The loss of fructose in the urine is smaller than that of glucose. Some authors claim that the tolerance for fructose locally in the infused veins is higher than that for the same concentration of glucose (*Thorén*, 1973b; *Job and Huber*, 1961).

c) Maltose. Young and Weser (1971) have proposed the use of maltose in parenteral nutrition. One molecule of this disaccharide is metabolised to two molecules of glucose. It therefore has half the osmotic activity of glucose and it should be possible to give a 10% maltose solution peripherally without damaging the vein more than a 5% glucose solution does. This is an obvious advantage. Maltose is also taken up by the body cells without insulin, but large amounts of maltose are excreted in the urine and may cause renal damage.

Kohri et al. (1972) found that maltose was metabolised as glucose in man and in guinea-pigs. The metabolic rates of the two carbohydrates showed a remarkable similarity. Where the rapid splitting of the maltose takes place is at present unknown. There is no maltase in the serum of man and guinea-pigs in contrast with other animals (*Kohri et al.*, 1972; *Yoshimura et al.*, 1973). 50% of infused maltose is rapidly metabolised to carbon dioxide – shown as $^{14}CO_2$ – and the other 50% assimilated in the body. Other Japanese authors have reported a very low toxicity of maltose. Unfortunately, clinical investigations in man have not yet clearly shown that maltose can be used, as a glucose substitute, to produce the desired effects on N utilisation and on body weight.

Foerster et al. (1975) have investigated the suitability of maltose for parenteral nutrition in healthy adults. A 20% infusion solution was used. Only when a low dosage (0.125 g/kg/h) was given, could a steady state in the level of maltose in the blood be obtained. The results showed that the humans have a limited capacity to metabolise maltose. About 20–30% of the infused maltose was lost in the urine. *Foerster et al.* (1975) concluded that maltose is not suited for intravenous nutrition.

3. Polyols

a) Sorbitol. Sorbitol is converted by sorbitol dehydrogenase to fructose, which is then metabolised in the way mentioned above. The enzyme sorbitol dehydrogenase in the liver is very active, even in a severely damaged organ (*Hoshi*, 1963). The loss of sorbitol in the urine – 4–25% according to *Principi et al.* (1973) – is higher than that of glucose or fructose. Investigations by *Keller and Froesch* (1972) have shown that only the conversion of sorbitol via fructose to glucose in the liver is independent of insulin. The further metabolism of the glucose is, however, only possible if insulin is available.

b) Xylitol. Xylitol is a pentitol and, consequently, enters directly into the metabolic pentose phosphate cycle. This has opened a new route of carbohydrate supply. At present, however, no situation is known where xylitol has any definite advantages over other carbohydrates, which in reality, means glucose.

Like fructose and sorbitol, most of the xylitol, 85% according to *Froesch and Keller* (1972) and *Müller et al.* (1967), is transformed into glucose. Insulin is therefore required for adipose tissue and muscles to build up triglycerides or glycogen, respectively, from xylitol, but as with fructose, liver glycogen is accumulated independently of insulin (*Froesch and Keller,* 1972). Xylitol also inhibits gluconeogenesis more efficiently than glucose (*Froesch,* 1972).

Schultis and Geser (1970) and *Bässler* (1970a, b) found xylitol more antiketogenic and more N-sparing than glucose. In several papers, *Schultis and Geser* (1970), *Schultis* (1971) and *Schultis and Beisbarth* (1972) studied the metabolism in post-operative conditions, and found improved N balance when xylitol was administered instead of glucose. *Halmagyi and Isvang* (1968) have treated 1,189 patients in the post-operative phase with xylitol, and found it to be equally well utilised both before and after operation.

Many authors (*Donahoe and Powers,* 1970; *Foerster et al.,* 1970; *Schumer,* 1971) have carefully studied the complications following xylitol infusions, such as marked hyperuricaemia, hyperbilirubinaemia, and renal insufficiency. The diuresis was later followed by anuria as well as liver and brain symptoms. The extensive investigations by *Thomas et al.* (1972) also revealed other symptoms (nausea and vomiting) and lethal lesions, such as centrilobular liver necrosis and liver enzyme abnormalities. A peculiar autopsy finding was that of calcium oxalate crystals in the renal tubuli and, in one case, in the brain. The signs of intoxication were proportional to the infused dose, but even the highest dose of 0.49 g/kg/h for, on the average, 40 h did not affect more than 20% of the patients. These symptoms appeared quite individually, and the authors made the reservation that the primary disease may have been responsible for the reactions. Later investigations by *Thomas et al.* (1975) have shown that intravenous infusions of xylitol in vitamin B_6-deficient rats produced very large urinary excretions of oxalate. The oxalate seems to be formed from glyoxylate, which occurs in large amounts during infusion of xylitol. It should, however, be pointed out that many workers believe that by giving xylitol simultaneously with fructose and glucose, the incidence of side-effects is reduced and the utilisation of the carbohydrate is improved (*Ahnefeld et al.,* 1975).

c) Glycerol. In dilute solutions and moderate dosage, glycerol is well tolerated in man and animals. In large doses and in solutions with a concentration of 10% or more, glycerol can cause haemolysis, hypotension, central nervous disturbances and convulsions (*Geyer,* 1960). These effects make glycerol impossible to use as the only energy supply in parenteral nutrition. In parenteral preparations, glycerol is thus used only in certain fat emulsions to correct their osmotic pressure. Being a three-carbon alcohol, it has twice the osmotic effects of monosaccharides. The calorific value of glycerol is almost the same as that of glucose.

4. Ethanol

Ethanol was originally suggested as an energy substrate in parenteral nutrition because of its high calorific value (7 kcal/g; 29 kJ/g). About 0.1 g ethanol will be metabolised/kg body weight/h. This corresponds to 2.4 g or 17 kcal (71 kJ)/kg/day. It has been shown that using alcohol as a part of the non-protein energy, a N-sparing effect can be obtained. After infusion, ethanol is converted in the liver to acetaldehyde and acetylcoenzyme A (acetyl-CoA), with NAD being reduced to NADH. However, the oxidation of lactate to pyruvate also requires the oxidised form of this coenzyme (fig. 2). The use of ethanol will thus increase the risk of lactic acidosis, particularly if fructose or sorbitol are being infused at the same time as ethanol (*Woods,* 1975).

The use of ethanol for intravenous nutrition is declining because of its pharmacological and toxic effects. With glucose and the modern fat emulsion, there should usually be no serious problems in supplying the required non-protein energy.

5. Conclusions

Fructose, sorbitol and xylitol are metabolised in the liver independently of insulin. The level of carbohydrate metabolites in the liver will then be increased. Most of the glucose produced in the liver originates from the given glucose substitutes and hence less amino acids are used for gluconeogenesis. A short-lasting N-sparing effect of fructose, sorbitol and xylitol will therefore be obtained. However, a protein anabolism in muscle and other peripheral tissues can

Table IX. Metabolic properties and some adverse reactions of carbohydrates and polyols for parenteral nutrition

	Carbohydrates		Polyols	
	glucose	fructose	sorbitol	xylitol
Normal metabolite	+	+		
Metabolised by cells in all tissues	+			
Increased nitrogen utilisation	+	+	+	+
Antiketogenic	+	+	+	+
'Partial insulin independency'		+	+	+
Metabolic acidosis		+	+	+
Hyperuricaemia		+	+	+
Decrease of adenine nucleotide in the liver and phosphate in the blood		+	+	+
Osmotic diuresis with electrolyte losses			+	+
Crystal deposition in kidney and brain				+

only occur when insulin will make glucose available to these tissue cells. The often repeated statement that the glucose substitutes — fructose, sorbitol and xylitol — are able to increase the glycogen stores in the liver independently of insulin, seems to be of limited value. In insulin resistance, the body has a reduced possibility of utilising the stored glycogen, which is thus restricted from the metabolism, when the need is most urgent. Some of the metabolic effects of glucose, fructose, sorbitol and xylitol are summarised in table IX.

The advantage of glucose over fructose, sorbitol and xylitol is obvious. All tissue cells can metabolise glucose, which is a prerequisite for anabolism. Glucose is required by the brain, stimulates insulin secretion and thereby promotes anabolic effects. However, the muscle and adipose tissue will assimilate glucose only in the presence of insulin. Exogenous insulin has thus to be given when endogenous insulin is not available.

It is important to remember that there are some potential hazards in the use of intravenous glucose. If patients with glucose 'intolerance' receive glucose as the sole non-protein energy source in parenteral nutrition, they might develop hyperglycaemia, with its potential sequelae of glucosuria and osmotic diuresis, dehydration and hyperosmolar coma. However, the blood glucose can be maintained below 11 mmol/l (200 mg%) and glucosuria kept at less than 2% by infusing glucose at a constant rate, and by giving a sufficient amount of exogenous insulin. A high endogenous insulin production stimulated by hypertonic glucose may cause a reactive hypoglycaemia if the glucose infusion is stopped suddenly. For this reason, Moore and Ball (1959) advised a 'chaser' of 5% glucose to follow hypertonic glucose infusion solutions.

One practical advantage which is worth emphasising about the use of glucose is that unlike all the other carbohydrates and polyols, the blood concentration of glucose can be estimated on an emergency basis, allowing rapid treatment of side-effects which usually are readily treatable.

Some nutritionists are of the opinion that fructose, sorbitol and xylitol are too dangerous to be used as energy source in parenteral nutrition. The main objections are the resulting acidosis, hyperuricaemia and depletion of adenine nucleotide in the liver and inorganic phosphorus in the blood (table IX). These complications occur in an unpredictable manner and cannot be controlled. Fructose or sorbitol should not be used in a paediatric clinic, in order to avoid fatal reactions in a non-identified case of fructose intolerance. Indeed, a decision has recently been made by the 'IKS' (Intercantonal Control Centre) prohibiting the use of fructose, sorbitol and xylitol for intravenous alimentation (*Fischer,* 1974).

When a solution of an amino acid mixture with carbohydrate as glucose or fructose is heat-sterilised, a destruction of the essential amino acids, such as lysine may occur ('Maillard-reaction'). This will result in a decrease in the biological value of the amino acid mixture. In order to avoid this reaction, sorbitol or xylitol have been used instead of glucose or fructose. However, using a modern technique for sterilisation, and by adjusting the pH of the solutions to less than 5.5, it is now possible to prepare infusion solutions containing a mixture of amino acids and carbohydrate without any Maillard-reaction occurring.

All investigations seem to indicate that *glucose is the carbohydrate of choice* for parenteral nutrition. Nonetheless, there are many authors who think that the special metabolic pathways of fructose, sorbitol, xylitol and, possibly, maltose make them of special importance in some clinical conditions. There can be no denying that fructose in particular has been successfully used as carbohydrate source on very many occasions. It is evident, however, that it is an illusion to regard these substances as 'insulin-independent'. It must be questioned whether energy substrates which must first be converted by the liver into glucose to be utilised in peripheral tissues have any advantages.

D. Fat

1. Some Biochemical Aspects on the Metabolism of Fat

The mechanism of absorption of orally ingested fat depends upon the length of the fatty acids involved. Short chain fatty acids produced by digestion of triglycerides pass directly into the portal vein, but long chain fatty acids (greater than 12 carbon atoms long) are re-esterified to triglycerides in the mucosal cells. They are then covered with a phospholipid and protein layer to form chylomicrons (*Frazer,* 1958). This delicate structure, where substances of decreasing hydrophilia occur from the particle surface to its centre, is the body's way of keeping the water-insoluble fat in a fine and stable emulsion in water. The chylomicrons, which are 1 μm or less in diameter, are carried by the thoracic duct to the subclavian vein and into the general circulation. This special way of transport is of interest for parenteral nutrition as in this case nature conveys the fat, including the fat-soluble vitamins A, D, E and K, directly into the venous blood in about the same manner as in intravenous nutrition with fat emulsions.

The chylomicrons are disintegrated in two different ways. There is an enzyme, lipoprotein lipase in various organs which hydrolyses the triglycerides of the chylomicrons. This reaction makes the free fatty acids (FFA) from the chylomicrons available for the tissues. As the chylomicrons disrupt, the milky lipaemic plasma rapidly clears, and hence lipoprotein lipase is also called 'the clearing factor'. This enzyme may be activated by heparin and released into the bloodstream (*Engelberg*, 1956).

The other way by which the post-prandial plasma is cleared, is mediated by the liver (*Olivecrona et al.*, 1961), where the triglycerides are split and some of the fatty acids liberated as FFA.

The ingested lipid may be used as a functional component of all cells, since lipids are important in many of the membrane-related activities of the cell. The bulk of the dietary lipid, however, is usually used as an energy source, either immediately in many of the body tissues, or after storage for a variable period of time in one of the fat depots.

A number of factors control the hydrolysis of the triglycerides (lipolysis) in adipose tissue. Powerful lipolytic agents are epinephrine, norepinephrine and human growth hormone, which all lead to activation of lipase via the adenyl cyclase-cyclic AMP mechanism (*Robison et al.*, 1971). The FFA, which are produced enter the circulation bound to serum albumin. The normal serum concentration of FFA is about 20 mg/100 ml (0.5–1.0 mmol/l) and this has a very rapid turnover rate of up to 30%/min. As 1 g of fat gives 37 kJ and the plasma volume is 3 litres, the maximum daily production of energy from FFA would be $0.200 \times 30/100 \times 3 \times 37 \times 0.001 \times 60 \times 24$ MJ, i.e. 10.0 MJ, corresponding to normal energy expenditure. With our normal diet, the plasma FFA are not usually metabolised at maximal speed, and about 50% of the energy needs are covered by carbohydrate.

Fatty acids are oxidised by a process of successive β-oxidations. Pairs of carbon atoms are removed until acetic acid is left. This acid, bound to CoA, either joins the Krebs' cycle to be completely oxidised, or unites in pairs to form the ketoacid acetoacetic acid (CH_3COCH_2COOH). Acetoacetic acid and other ketone bodies, β-hydroxybutyric acid and acetone, formed by the liver, circulate in the blood and are oxidised by various tissues to carbon dioxide, water and energy.

Since the intramitochondrial level of oxaloacetic acid is maintained by carboxylation of pyruvic acid, on a high fat and low carbohydrate diet, an

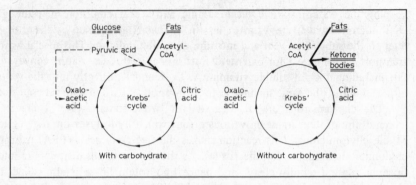

Fig. 3. The biochemical basis of 'fat burns in the fire of carbohydrate'.

insufficient amount of oxaloacetic acid may be available for the oxidation of the acetyl-CoA. The accumulating acetyl-CoA forms a surplus of ketone bodies, and ketosis develops. This is the modern biochemical explanation of the phrase, that 'fat can only be burnt in the fire of carbohydrate' (fig. 3). For optimal metabolism of fat it is therefore clear that an adequate carbohydrate intake should also be provided.

Glycerol released by hydrolysis of triglycerides is transported to the liver where it is metabolised along the same pathways as carbohydrate. It is converted into triosephosphate, pyruvate and, finally, oxidised through the tricarboxylic acid cycle. However, the energy yield from oxidation of glycerol is very small compared with that from oxidation of the fatty acid components of the triglyceride. For example, from glycerol tripalmitate, the complete oxidation of the triglyceride leads to a net production of 410 molecules of ATP, 390 of which are derived from the 3 palmitic acid molecules and only 20 from the glycerol.

2. Fat Emulsion for Intravenous Nutrition

a) Chemical and physical properties of fat emulsions. A fat emulsion for intravenous nutrition contains a vegetable oil in water and emulsifiers to stabilise the emulsion. Three commercial fat emulsions are *Intralipid, Lipiphysan* and *Lipofundin S.* The composition of these emulsions is given in table X. They contain either cottonseed oil or soybean oil. In Intralipid, purified egg yolk phospholipids are used as emulsifiers, whilst in the other emulsions soybean phospholipids or lecithin are used. Isotonicity with the blood is obtained by the addition of sorbitol, xylitol or glycerol.

In addition, two emulsions are available which also contain added amino acids. Nutrifundin and Trivémil both contain 38 g/l soybean oil, 60 g/l amino acids, and 100 g/l polyols (table XI).

Table X. Composition of intravenous fat emulsions

	Intralipid[1]	Lipiphysan[2]	Lipofundin[3,4]	Lipofundin S[3]
Soybean oil, g	100 or 200			100 or 200
Cottonseed oil, g		150	100	
Egg yolk phospholipids, g	12			
Soybean lecithin, g		20		
Soybean phospholipids, g			7.5	7.5 or 15
Glycerol, g	25			
Sorbitol, g		50	50	
Xylitol, g				50
DL-α-Tocopherol, g		0.5	0.585	
Distilled water to a volume of, ml	1,000	1,000	1,000	1,000

[1] Vitrum, Stockholm. [2] Egic, Loiret. [3] Braun, Melsungen. [4] Lipofundin is no longer commercially available.

Intralipid is the most widely used fat emulsion. The fatty acid patterns of the soybean oil and egg-yolk phospholipid in Intralipid are shown in table XII.

Schoefl (1968) and *Fraser and Håkansson* (1973) have made a series of electron microscopic studies on chylomicrons and Intralipid. These investigations showed that the particle size in Intralipid 10% was about the same as that of chylomicrons (table XIII). The particles in Intralipid 20% were found to be somewhat larger. According to *Fraser and Håkansson* (1973), the surface area of the fat particle in Intralipid 10% covered by one molecule of phospholipid was $23.9-37.4 \times 10^{-8} \mu m^2$. The corresponding area on chylomicrons was $60.9-72.5 \times 10^{-8} \mu m^2$. These studies and other investigations have shown pronounced physical similarities between natural chylomicrons and the fat particles of the fat emulsion Intralipid.

b) Pharmacological properties of fat emulsions: studies on animals. Since in clinical practice fat emulsions are often given over a long period of time, it is of great importance to show that during their *long-term administration* no toxic effects are produced. This is of particular relevance since the early cottonseed oil emulsion preparation Lipomul had to be withdrawn from use because of the incidence of side-effects. Chronic toxicity tests of modern fat emulsions have been made on dogs in many laboratories.

Figure 4 illustrates the results in 40 dogs given fat emulsions in amounts of 9 g fat/kg/day for 4 weeks (*Jacobson and Wretlind,* 1970). One of the fat emulsions was Intralipid, containing soybean oil emulsified with egg yolk phospholipids. Also, a soybean oil emulsion stabilised with soybean phospholipids

Table XI. Fat-polyol-amino acid preparations

Constituents	Nutrifundin[1]	Trivémil[2]
Soybean oil fract., g	38	38
Soybean lecithin, g		7
Soybean phospholipids fract., g	3.8	
Sorbitol, g		100
Xylitol, g	100	
Isoleucine, g	3.3	3.3
Leucine, g	5.8	6
Lysine, g	3.6	2.4
Methionine, g	2.7	2.7
Phenylalanine, g	3.3	3.3
Threonine, g	2.65	2.7
Tryptophan, g	1.1	1
Valine, g	3.15	3
Alanine, g	10.7	14
Arginine, g	–	1
Asparagine, g	2.1	–
Glycine, g	7.25	10.75
Histidine, g	–	1
Ornithine, g	6.1	1
Proline, g	5.85	5.75
Serine, g	1.6	1.6
Tyrosine, g	0.8	0.5
Distilled water to, ml	1,000	1,000

[1] Braun, Melsungen. [2] Egic, Loiret.

(Lipofundin S) was tested. Two cottonseed oil emulsions, Lipiphysan and Lipofundin, were investigated concurrently with Intralipid in the same doses and with the same experimental technique.

All the animals on Intralipid survived and gained weight, while all the dogs given the 3 other fat emulsions died. The Intralipid-treated animals showed only a moderate degree of anaemia and leucocytosis. The group of dogs on cottonseed oil and Lipofundin S showed marked leucocytosis and anaemia, as well as hypertriglyceridaemia, vomiting, diarrhoea, and blood in the urine and faeces. In similar studies performed by *Obel* (1970), the cause of death was found to be widespread fat embolism. The animal tests confirmed the severe toxic effects of cottonseed oil-soybean phospholipid emulsions found in earlier studies by

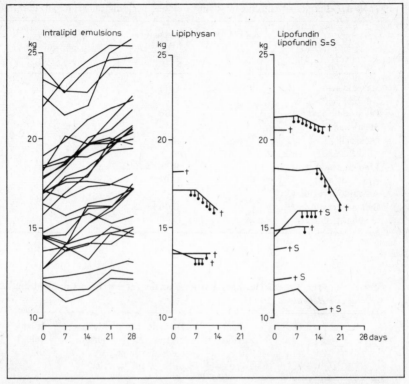

Fig. 4. A comparison of intravenous fat emulsions. Body weight and toxic reactions in dogs given fat emulsions intravenously (dose: 9 g fat/kg/day) for planned 28 consecutive days. Each line corresponds to the progress of one dog. † Signifies death, and ♦ signifies bleeding from the gastrointestinal or urinary tract.

Håkansson (1968). According to *Voss and Schnell* (1970), Nutrifundin causes no adverse reactions in dogs.

The administration of intravenous fat as Intralipid to rats at the dose of 9 g fat/kg/day, 5 days a week, for 4 weeks has been found to have no effect on reproduction nor was there any adverse effect found in the progeny of in-jected rats.

c) Elimination of fat particles from the bloodstream. The elimination kinetics from the bloodstream of intravenously injected Intralipid and natural chylomicrons were studied by *Carlson and Hallberg* (1963) in fasting dogs. They found the same rate of elimination for both Intralipid and chylomicrons.

Hallberg (1965a, b) showed later that the kinetic principles for the elimina-tion of chylomicrons and the fat emulsion Intralipid from the bloodstream were

Table XII. Fatty acid patterns of soybean oil and egg yolk phospholipids in Intralipid

		Soybean oil %	Egg yolk phospholipids %
Myristic acid	C_{14}	0.04	0.09
Palmitic acid	C_{16}	9	33
Palmitoleic acid	$C_{16:1}$	0.03	0.4
Stearic acid	C_{18}	3	16
Oleic acid	$C_{18:1}$	26	32
Linoleic acid	$C_{18:2}$	54	11
Linolenic acid	$C_{18:3}$	8	0.3
Arachidic acid	C_{20}	0.1	0.1
Arachidonic acid	$C_{20:4}$	–	0.2
Behenic acid	C_{22}	0.06	3
Unidentified acids		–	0.2

Table XIII. The size of the lipid particles in the fat emulsion Intralipid and chyle (*Fraser and Håkansson*, 1973)

	Mean diameter μm	Diameter of particles at median volume μm	Mean surface area μm^2	Mean volume μm^3
Artificial emulsions				
Intralipid 10%	0.13	0.32	0.092	4.54×10^{-3}
Intralipid 20%	0.16	0.43	0.13	8.38×10^{-3}
Natural emulsions				
Rabbit chyle, 5% corn oil diet	0.096	0.14	0.029	0.65×10^{-3}
Rabbit chyle, 30% corn oil diet	0.15	0.27	0.067	1.89×10^{-3}
Rat chyle, at peak of fat absorption	0.21	0.36	0.17	8.99×10^{-3}

also similar in man. These principles are as follows: at a high concentration of the infused lipids in the bloodstream, the elimination process is maximal and proceeds at a constant rate down to a so-called critical concentration. Below this concentration, the rate of elimination is dependent on the concentration, i.e. a fractional removal rate.

From studies in adults, on the rate of elimination from the blood of infused Intralipid, *Hallberg* (1965a, b) found, in subjects fasting overnight, that 3.8 g fat/kg body weight/24 h was eliminated from the bloodstream. This amount corresponds to 35 kcal or 145 kJ/kg/24 h. After fasting for 38 h, the elimination capacity increased to 52 kcal (218 kJ)/kg/24 h. In the post-operative period, after fasting for about 2 days, the elimination capacity was equivalent to 100 kcal (420 kJ)/kg/24 h. *Wilmore et al.* (1973a, b) have also confirmed this increased fat elimination capacity in the excessively catabolic state of patients suffering from burns. *Forget et al.* (1974) have shown an increase in the fat elimination rate when the intravenous supply of Intralipid was successively increased from 3 to 8 g fat/kg/day in a child. At the same time, an increase in the post-heparin lipoprotein lipase activity was observed.

Rössner (1974) has developed a clinical test for fat elimination kinetics. He points out that there are great individual variations in this respect. Besides differences in the nutritional state, the elimination rate is the integrated result of the action of several hormones. It is clear that there are major differences in elimination rate of different parenteral fat emulsions. According to *Lemperle et al.* (1970), the elimination of the fat from the bloodstream, after infusions of the fat emulsions Lipiphysan and Lipofundin S, is much more rapid than after Intralipid, and thus also more rapid than the chylomicron rate. The optimal elimination rate of infused fat from the bloodstream is not known. From a physiological point of view, however, the elimination rate of chylomicrons may be regarded as optimal.

Coran and Nesbakken (1969) compared the effect of intravenously administered Intralipid 20%, on plasma triglycerides and FFA in adult dogs and 5 new-born puppies. They concluded that the new-born animal is capable of metabolising intravenously administered fat just as readily as the adult.

d) Fat emulsions and reticuloendothelial system (RES). Scholler (1968) showed that the fat from some fat emulsions is accumulated in the cells of the RES. This accumulation depended on the composition of the emulsions and the size of the fat particles. After intravenous infusions of Lipofundin and Lipofundin S, the fat particles were taken up by the RES cells by phagocytosis in both man and experimental animals (*Lemperle et al.,* 1970). The RES was partly blocked and antibody formation was significantly reduced. After single infusions of Intralipid in guinea-pigs, however, no accumulation of the fat particles in the Kupffer cells was observed and there was also no significant reduction in the formation of antibodies (*Lemperle et al.,* 1970). Since body resistance to infections is impaired by the accumulation of particles in the RES, *Scholler* (1968) stated that only those fat emulsions should be used clinically which, like the natural chylomicrons, are not taken up by the RES.

Meyer et al. (1957) observed a brown pigment in the RES of patients and dogs treated repeatedly with cottonseed oil infusions. Intravenous fat pigment in the RE cells has been observed also after soybean oil emulsions, but to a much lesser extent than with cottonseed oil emulsions (*Thompson et al.,* 1965). The appearance of pigment in the liver and spleen of both man and animals after long-term fat infusions is interesting from a pathological point of view. Numerous liver-function tests were performed (*Thompson et al.,* 1965) in an attempt to assess the potential hazards which might be associated with large doses of emulsions, such as the deposition of intravenous fat pigments in the liver, the development of proliferative lesions or alterations within hepatocytes. No impairment of liver function was observed either during the course of infusion or afterwards.

Groups of rats have been given 3, 6 and 9 g fat as Intralipid/kg/day for 4 weeks and then sacrificed after various periods up to 1 year. The haematological values of all the rats were normal. Histology of all major organs was performed. Particular attention was directed to intravenous fat pigment and proliferation of the RE cells (microgranulomas) in the livers and spleens. It was found that the intravenous fat pigment was present immediately after the injection period. The deposition of pigment increased and peaked at 3 months after the injections. During the subsequent months, the pigment decreased. Microgranuloma formation was practically non-existent immediately after the injections and at 12 months after. Although very infrequent, the maximum incidence of microgranulomas appeared to be at about 1 month after the injection period.

e) Fat emulsions and coagulation system. Investigations by *Cronberg and Nilsson* (1967) as well as *Reid and Ingram* (1967) showed that infusion of Intralipid has no effect on coagulation or on the fibrinolytic system.

Kapp et al. (1971) investigated the influence of the Intralipid infusion on platelet adhesiveness and platelet count. There was a significant reduction in platelet adhesiveness, averaging 15% during infusion of 500 ml Intralipid 20%. The initial degree of adhesiveness was re-established, to a considerable extent, 4 h after termination of the infusion. The platelet count did not change significantly during Intralipid infusion. Control infusion with a physiological saline solution, or a 5% glucose solution, produced no significant change in adhesiveness.

The fact that thrombophlebitis occurs very rarely, when Intralipid fat emulsion is infused into a peripheral vein, may be regarded as a clinical sign that this emulsion does not have any major influence on the coagulation system.

Infusion of cotton seed oil emulsion causes hypercoagulation in rabbits (*Huth et al.,* 1967). After Intralipid infusion no significant changes were noted.

f) Fat emulsion and pulmonary function. *Steinbereithner and Wagner* (1967) investigated the effect of rapid intravenous infusions of Intralipid on the arterial oxygen tension in patients with serious cranial trauma and chronic hypoxia. The results of these studies showed no reduction of the oxygen tension and no interference with oxygen supply to the tissues.

Sundström et al. (1973) have examined some pulmonary function parameters before, during and after infusion of Intralipid. The pulmonary diffusion capacity for carbon monoxide decreased about 15% with the infusion, but recovered to the basal value 45 min after the end of infusion. The alveolar-arterial difference in oxygen tension decreased with the Intralipid infusion. There was no change in the alveolar oxygen tension, but an increase in the arterial oxygen tension.

The pulmonary diffusion capacity in patients with burns, determined by [133]Xenon perfusion-diffusion, and carbon monoxide rebreathing technique, was found to be normal after the infusion of Intralipid (*Wilmore et al.,* 1973a, b). Blood gas levels did not change after infusion of single or multiple units of the fat emulsion.

Measurement of steady state pulmonary diffusion capacity and membrane-diffusing capacity in normal human volunteers, before and after administration of 500 ml of Intralipid 10% intravenously, demonstrated a significant decrease in those functions for at least 4 h following the infusion in 6 of 10 subjects (*Greene et al.,* 1976). These changes returned to control levels within 24 h in all 6 subjects. Simultaneous infusion of heparin with Intralipid prevented the decrease in both the pulmonary diffusion capacity and the membrane diffusing capacity. This finding may support the concept of adding heparin to Intralipid post-operatively, 2,500 IU/500 ml of Intralipid 20% (*Jacobson,* 1974), particularly when pulmonary function is abnormal.

g) Importance of essential fatty acids. Burr and Burr (1930) discovered that certain polyunsaturated fatty acids were essential for growth and survival of rats. The fatty acids which cannot be produced in the body were found to be linoleic acid and linolenic acid, with 2 and 3 double bonds, respectively. Arachidonic acid with 4 double bonds, which is also necessary for normal metabolism, can be produced in the body from linoleic acid.

That essential fatty acid deficiency may lead to a characteristic dermatitis was first suggested by *Hansen* (1933). With the use of fat-free parenteral nutrition in the United States, a number of cases have now been reported in infants. A typical case was documented by *Caldwell et al.* (1973). A 25-week-old baby on parenteral nutrition, but with no fat emulsion since its 18th day of life developed scaly skin lesions, sparse hair growth, and thrombocytopenia. Analysis of the fatty acids in the plasma phospholipids showed low levels of linoleic and arachidonic acids, and a high content of 5,8,11-eicosatrienoic acid, indicating an essential fatty acid deficiency. With the administration of Intralipid in a quantity sufficient to provide 4% of the daily energy requirement as linoleic acid, the levels of linoleic and arachidonic acid rose, the level of 5,8,11-eicosatrienoic acid decreased, the skin lesions healed, and the thrombocytopenia was corrected.

The above changes in the fatty acid content of plasma phospholipids have been detected within a few days of commencement of fat-free intravenous nutrition in infants (*Paulsrud et al.,* 1972). For a long time, there were no definite indications that the essential fatty acids were required by adults. An adult male patient, however, who was maintained on intravenous nutrition without fat for 100 days, developed signs of essential fatty acid deficiency (*Collins et al.,* 1971). Clinically, he had a characteristic skin rash, and biochemically the serum phospholipids were found to contain 10% of 5,8,11-eicosatrienoic acid with a low content of arachidonic acid. Intralipid administered intravenously (23 g linoleic acid/day) caused the content of eicosatrienoic acid in the serum phospholipid to fall and arachidonic acid to rise. Simultaneously, the serum triglycerides fell to normal levels, and the rash disappeared.

In 5 of 13 adult patients with severe burns, it was found that the essential, polyunsaturated acids, as well as phosphatidyl serine and phosphatidyl ethanolamine, were much reduced in the erythrocyte membranes (*Wilmore et al.,* 1973b). The changes were not correlated with the degrees of the burns. Stress or alcoholism were mentioned as possible additional aetiological factors. Fat emulsion (Intralipid) given parenterally restored the lipid composition of the erythrocyte membrane.

Essential fatty acids are also required to maintain a normal lipid composition in other tissues, e.g. brain lipids (*Galli et al.,* 1970). *Jeejeebhoy et al.* (1973) found that fat infusions were necessary to prevent the development of fatty liver in a patient receiving complete parenteral nutrition.

Both linoleic (ω 6) and linolenic acid (ω 3) exhibit essential fatty acid activity. However, the two acids are not physiologically interconvertible. *Rivers and Davidson* (1974) found that second-generation mice fed on a diet containing linoleic and linolenic acid had a lower fasting metabolic rate than mice given a linolenic acid deficient diet. *Sinclair* (1974) showed that the addition of linseed oil containing linolenic acid could cure skin lesions occurring in monkeys on a diet rich in linoleic acid. Histologically, the fatty infiltration of the liver in the monkeys also disappeared when linolenic acid was added to the linoleic acid-rich diet. The total lipid in the liver was reduced from a level between 300 and 600 g, to a level between 190 and 240 g/kg dry matter. Both these investigations seem to indicate that linolenic acid is essential.

The importance of essential fatty acids in nutrition has recently been emphasised by the demonstration that the prostaglandins are synthesised from linoleic acid, via arachidonic acid. One of the effects of the prostaglandins is an antilipolytic action, which results from the inhibition of adenyl cyclase. The prostaglandins also prevent aggregates of platelets and, possibly, thrombus formation.

The thrombosis tendency *in vivo* can be reduced by increasing the polyunsaturated fatty acid intake at the expense of saturated fatty acids (*Hornstra,* 1974). Linoleic acid and, possibly, linolenic acid, appear to be much more active in this respect than oleic acid.

The amount of essential fatty acids required to prevent the development of deficiency symptoms has been the subject of some debate. According to *Collins et al.* (1971) about 2–7 g linoleic acid/day (0.1 g/kg body weight/day) is required in adults on intravenous nutrition. *Jeejeebhoy et al.* (1973), however, have suggested that 25 g linoleic acid/day may be required. On the other hand, *Press et al.* (1974) found that 230 mg sunflower oil/day, rubbed into the flexor surface of the forearm, was capable of increasing the serum level of linoleic acid and decreasing that of eicosatrienoic acid. Whether this type of treatment has any effect on tissue fatty acids remains to be seen. At present, it can be concluded that the daily infusion of 500 ml 10% soybean oil emulsion, which contains about 27 g linoleic acid and about 4 g linolenic acid, should satisfy the requirements of all adult patients. In infants, infusion of 2–4 g fat emulsion/kg/day will provide adequate amounts of essential fatty acids.

3. Utilisation and Adverse Reactions of Fat Emulsion in Man

a) Utilisation. Investigations have shown that fat emulsions are utilised in a similar way to fat derived from ordinary food. Fat does not cause diuresis. No losses of fat are observed either in the urine or in the faeces. With isotope techniques, it has been observed that fat emulsions administered intravenously are rapidly metabolised (*Geyer et al.,* 1948; *Eckart et al.,* 1973/74). An intravenous infusion technique in rats has been used to determine the oxidation rate of the fatty acids in fat emulsion (*Lindmark and Wretlind,* 1975; *Yokoyama et al.,* 1975). Soybean oil emulsion (Intralipid) labelled with ^{14}C palmitic, oleic or linoleic acid has been used. The peak of $^{14}CO_2$ in the expiration air, indicating maximum oxidation of the fat, occurred 2 h after the injection. The total amount of fat oxidised during the first 12 h after the injection was 40–70% depending on – among many different factors – the nutritional state of the rat. *Reid* (1967) has reported a decrease in the RQ-value, indicating increased fat metabolism after intravenous infusion of fat emulsions. A positive energy balance obtained by intravenous supply of fat emulsions produced the calculated weight increase (*Jacobson and Wretlind,* 1970). This finding also indicates a normal utilisation of the fat given. There are great biological differences, however, between various fat emulsions. Some of these have been summarised by *Wretlind* (1972).

Many studies indicate that fat emulsions can be used as a source of energy to improve N balance (*Schärli,* 1965; *Wadström and Wiklund,* 1964; *Reid,* 1967). Moreover, as has already been discussed (sect. IV.B.7.c), a number of comparisons have been made of intravenous regimes in which non-protein energy was supplied by carbohydrate alone or variable proportions of fat and carbohydrate. Although in the short-term carbohydrate alone is somewhat more protein-sparing, overall effects of different isocaloric regimes are very similar, testifying to the efficient utilisation of the fat emulsion.

Table XIV. Some reports on the use of Intralipid in adults during the years from 1961 to 1975

Fat/kg/day g	Infusion period	Authors	Years
1–3	Single infusions up to 28 days	*Schuberth and Wretlind* (1961) *Buchner and Cesnik* (1962)	1961–1962
1–2	Single infusions up to 41 days	*Østergaard* (1963) *Schuberth* (1963) *Freuchen* (1963) *Jones et al.* (1964) *Schärli* (1964)	1963–1964
1–3	Single infusions up to 150 days	*Lawson* (1965) *Lee and Shortle* (1965) *Hadfield* (1966) *Steinbereithner* (1966) *Peaston* (1966) *Hallberg et al.* (1966) *Freuchen and Østergaard* (1967) *Hartmann* (1967) *Vlaardingerbroek* (1967)	1965–1967
2–5	Up to 7 months	*Hartmann* (1968) *Feischl and Hiotakis* (1969) *Stell* (1970) *Liljedahl* (1970) *Jacobson et al.* (1971) *K. Bergström et al.* (1972) *Jacobson* (1972a, b) *Zumtobel and Zehle* (1972) *Geremy et al.* (1972)	1968–1972
2–5	Up to more than 7 years	*Jeejeebhoy et al.* (1973) *Wilmore et al.* (1973a) *Solassol and Joyeux* (1974) *Deitel and Kaminsky* (1974) *Lamke et al.* (1974) *Kessler* (1974)	1973–1975

In the somewhat artificial situation of a zero N intake in healthy volunteers, *Brennan et al.* (1975) showed that the sparing of N with infusion of fat emulsion (Intralipid) was smaller than with isocaloric glucose infusions. The conservation of N produced by the fat infusion could be duplicated by the infusion of glycerol alone in the same amounts as that available from the fat emulsion.

That it is possible to administer complete intravenous nutrition with fat in the form of Intralipid, for a long period in adult man, has been demonstrated in a large number of investigations (table XIV).

Lawson (1965) gave about 3 g fat/kg for a period varying between 8 and 36 days. Liver biopsies, after 30 and 36 days, showed a pigmentation of the Kupffer cells as the only change. No appreciable changes were observed in the brom-sulphthalein retention, or serum bilirubin or transaminases.

In one case, complete intravenous nutrition was administered for 7 months and 13 days (*K. Bergström et al.,* 1972). For 6 months, 408 ml of Intralipid 20% were given daily, corresponding to 816 kcal. The increase in weight during the period of intravenous nutrition was 10 kg, without any appreciable change in the body fluid volume. Histopathological studies were conducted by means of repeated needle biopsies of the liver (*Jacobson et al.,* 1971). The parenchymal cells did not show any significant light or electron microscopic changes during the period of intravenous nutrition. The Kupffer cells, however, showed focal proliferations, enlargement, accumulation of fat droplets and occurrence of a lipofuscin-like pigment. After cessation of the intravenous therapy, there was a slow decrease in the number of lipofuscin-like bodies in the Kupffer cells during the following 1.5 years. The observed changes did not signify any liver cell damage.

Jeejeebhoy et al. (1973) have a patient (born 1934) who, since 6 October 1970, has been exclusively nourished intravenously. The total energy supply has been 2,000 kcal or 34 kcal/kg. The daily supply of fat has been 50 g (500 ml Intralipid 10%). Two liver biopsies were performed after fat had been supplied for 3 and 5 months, respectively. They showed that the hepatic tissue was normal.

Complete intravenous nutrition with fat as Intralipid, has been thoroughly studied in neonates and infants (table XV). To neonates and infants on intra-venous nutrition, *Rickham* (1967) gave 2.5–3 g fat/kg body weight/day as Intralipid. *Børresen and Knutrud* (1969) and *Grotte* (1971) have given 3–4 g fat (Intralipid)/kg/day to neonates and infants, as part of a regime resulting in good growth.

A large number of children who had been treated with intravenous nutrition were given 28–69% of the energy supplied in the form of fat (Intralipid). The daily supply of fat was 1.1–6 g/kg. The infusions, given via a peripheral vein, have lasted up to 122 days. In the majority of all these seriously ill children, satisfactory growth was obtained.

There are still relatively few reports of clinical trials involving the emulsions containing amino acids. *Witzel et al.* (1970) have reported on the safe use of Nutrifundin in 23 patients treated with 1,003 infusions. However, *Wolf et al.* (1973) found that Nutrifundin did not support growth or reduce oedema in premature infants. As regards Trivémil, *du Cailar et al.* (1974) found this

Table XV. Some reports on the use of Intralipid in neonates and infants during the years from 1967 to 1975

Fat/kg/day, g	Infusion period	Authors	Years
2.5–4	3–50 days	Rickham (1967) Børresen et al. (1970a, b)	1967–1970
1–4	3–84 days	Coran (1972) Caldwell et al. (1973) Wei et al. (1972) Børresen (1972)	1971–1973
3–8	3–205 days	Grotte et al. (1974) Forget et al. (1974) Coran (1974) Pendray (1974) Puri et al. (1975) Schärli and Rumlova (1975)	1974–1975

product to be well tolerated in patients at a dose of 1,000–1,500 ml daily. Many investigators have reported that they have found Trivémil to be non-toxic and valuable. However, more prolonged and carefully defined studies are required to ascertain its overall safety and usefulness.

    b) Adverse reactions. Acute adverse reactions have been observed only in a small percentage of the patients given the soybean oil-egg yolk phospholipid emulsion Intralipid. These reactions were febrile response (Hartmann, 1967), chills, sensation of warmth, shivering, vomiting, pain in chest and back, and thrombophlebitis (Hallberg et al., 1967). Five severe reactions were reported in paediatric patients (Chaptal et al., 1964, 1965). In most patients, this involved hyperlipaemia and hyperthermia. After 14–15 days of Intralipid infusion, anaemia was also observed in 2 children (Hanc et al., 1968; Chaptal et al., 1965). Krediet and De Gens (1975) described a 29-year-old woman on parenteral nutrition for 10 weeks because of cachexia and a large defect of the abdominal wall after a colectomy. After 47 days, during which she had received 4.2 kg fat as Intralipid, high fever, jaundice and thrombocytopenia occurred. These symptoms disappeared when the infusion of fat emulsion was discontinued. The question regarding the use of fat emulsions in patients with liver disease is discussed in section VIII.F.

    The administration of Intralipid is contraindicated in patients with severe disturbances of fat metabolism, such as pathologic hyperlipaemia and lipoid nephrosis (Freund et al., 1975).

    In order to prevent the supply of fat to a patient unable to eliminate the fat from the bloodstream, the following simple test should be performed. In the morning following the first day's infusion of fat emulsion, a citrated blood sample is centrifuged at 1,200–1,500 rpm. If the plasma is then strongly opalescent or milky, further infusion should be postponed or smaller daily amounts of fat emulsion should be given. In the very great majority of cases, plasma is completely clear 12 h after the conclusion of an infusion of fat. This test should be repeated at weekly intervals. It is extremely rare to find a patient who will not be able to eliminate the fat particles from the circulation.

4. Conclusions

Modern fat emulsions are free from the major adverse reactions of the earlier fat emulsions containing cottonseed oil. During the last decade, considerable experience in parenteral nutrition with fat emulsions has been gained.

With intravenous fat emulsions – in combination with carbohydrates – it is relatively easy to supply the required daily amount of energy. The main advantage of the fat emulsions is that a large amount of energy can be given in a small volume of isotonic fluid through a peripheral vein in contrast with the concentrated glucose solutions, which have to be given through a central vein catheter. Thrombophlebitis occurs very infrequently when isotonic fat emulsions are infused into peripheral veins, nor do they cause a hyperosmolar syndrome or diuresis. Furthermore, no losses are observed either in the urine or faeces. Fat emulsions supply the body with essential fatty acids and triglycerides, which are a part of ordinary food. In this way, the usual supply of fat may be maintained, and the lipid composition of the body kept normal.

However, *because there are pronounced differences in tolerance and toxicity between the various fat emulsions, it is incorrect to speak of fat emulsions for intravenous nutrition in general terms. The product's name, and exact composition should always be stated.* A soybean oil-egg yolk phospholipid emulsion has been shown to be very well tolerated both by animals and man. It seems to have the same properties as chylomicrons, and to be utilised in the same way as the fat from the ordinary diet. This preparation is therefore the fat emulsion of choice at the present time.

E. Minerals

Knowledge of mineral requirements is somewhat inadequate. Informative determinations of mineral balance have relatively seldom been made. Consequently, exact figures are not readily available for the normal state. This applies still more to disease, where the balances may be disturbed. Mineral metabolism is also apparently a more individual problem than the metabolism of amino acids, carbohydrates and fat described in the earlier part of this chapter (*Shils,* 1972). Table XVI summarises some recommendations of the electrolytes to be supplied in various situations when intravenous nutrition is indicated. A review of the

Table XVI. Tentatively recommended daily allowances of water, energy, amino acids, carbohydrates, fat, and minerals for patients on complete intravenous nutrition. The basal allowances will cover resting metabolism, some physical activity and specific dynamic action, but no increased needs because of trauma, burns, etc. The *moderate* amounts should be used when the patient has increased losses or is in a depleted status. The *high* supply should be used in severe catabolic conditions as burns, after trauma, etc.

	Allowances/kg body weight to adult			Allowances/kg body weight for neonates and infants		
	basal amounts[1]	moderate amounts	high supply	basal amounts[1]	moderate amounts	high supply
Water, ml	30	50	100–150	100–200	125	125–150
Energy	30 kcal = 0.13 MJ	35–40 kcal[3] = 0.15–0.17 MJ	50[9]–60[3] kcal = 0.21–0.25 MJ	90–120 kcal = 0.38–0.50 MJ	125 kcal = 0.52 MJ	125–150 kcal = 0.52–0.63 MJ
Amino acid	90 mg (0.7 g amino acids)	0.2–0.3 g (1.5–2 g amino acids)	0.4–0.5 g (3–3.5 g amino acids)	0.3 g (2.5 g amino acids)	0.45 g (3.5 g amino acids)	0.5 g (4.0 g amino acids)
Glucose, g	2	5	7	12–18	18–25	25–30
Fat, g	2	3	3–4	4	4–6	6
Sodium, mmol	1–1.4	2–3[4]	3–4[10]	1–2.5	3–4	4–5
Potassium, mmol	0.7–0.9	2	3–4[10]	2	2–3	4–5
Calcium, mmol	0.11	0.15	0.2[10]	0.5	1[19]	1.5–2
Magnesium, mmol	0.04	0.15–0.20[4]	0.3–0.4[10]	0.15	0.15–0.5[13]	1[14]
Iron, μmol	0.25–1.0	1.0[5]	1.0[5]	2		3–4
Manganese, μmol	0.1	0.3	0.6[1]	0.3[12]	0.7[14,15]	1[1]
Zinc, μmol	0.7[2]	0.7–1.5[5,6]	1.5–3[5,6]	0.6	1	1.5[5]
Copper, μmol	0.07	0.3–0.4	0.4–1[11]	0.3		
Chromium, μmol	0.015[2]			0.01[11]		
Selenium, μmol	0.006[2]			0.04[12]		
Molybdenum, μmol	0.003[2]					
Chlorine, mmol	1.3–1.9	2–3	3–4[10]	2–4	1.3[17]–1.5[18]	4[15]–6[10,16]
Phosphorus, mmol	0.15	0.4[7]	0.6[10]–1.0[5]	0.4–0.8		2.5[15]–3[5]
Fluorine, μmol	0.7	0.7–1.5[8]		3		
Iodine, μmol	0.015			0.04		0.1[15]

[1] *Wretlind* (1972). [2] *Jacobson and Wester* (1977). [3] *Bozzetti* (1976). [4] *Jeejeebhoy et al.* (1976). [5] *Seeling et al.* (1975). [6] *Weisman et al.* (1976). [7] *Jacobson* (1976). [8] *Shils* (1972). [9] *Deligné et al.* (1974). [10] *Giovanoni* (1976). [11] *Shils* (1974). [12] *Ricour et al.* (1975a). [13] *Børresen et al.* (1970a). [14] *Wilmore et al.* (1969). [15] *Dudrick et al.* (1976). [16] *Beyreiss* (1975). [17] *Ricour et al.* (1975b). [18] *Børresen* (1974). [19] Recommended Dietary Allowances (1968).

requirements and metabolism of minerals under basal conditions has been given earlier (*Wretlind,* 1972). It is evident that in situations of abnormal losses of minerals, whether due to loss of body fluids, e.g. intestinal secretions, exudates, etc., or to excess breakdown of body tissues as in hypercatabolic states, it is almost impossible to predict mineral requirements. The best estimate can be obtained by careful measurement of losses and monitoring the effect of replacement on blood concentrations of those substances which can be routinely analysed.

1. Sodium

There is a total of about 4,000 mmol (90 g) of sodium in the body, the bulk of which is extracellular, and it is therefore of special importance for the maintenance of the volume of the extracellular fluid and hence also the circulating blood volume. The supply of sodium to cover the basal losses has been estimated to amounts between 1 and 1.4 mmol/kg/day (table XV). 90% of ingested sodium is eliminated in the urine, and only 10% in sweat and faeces under normal conditions. The overall balance of sodium is well regulated and the normal body has a considerable capacity for protecting itself against sodium deficiency. The kidneys are able to limit the loss of sodium to about 1 mmol/l of urine when the intake is small or zero. The adrenal cortex is the main regulatory organ through the action of aldosterone, but it is still not known to what extent its function is disturbed in disease and other forms of stress.

In many cases on intravenous nutrition, large quantities of sodium have been given. *Dudrick et al.* (1969) gave between 125 and 160 mmol (2–3 mmol/ kg) sodium/day. Where alimentation without fat (hyperalimentation) is applied, an osmotic diuresis often occurs, leading to losses of electrolytes, and addition of extra sodium chloride is necessary.

For neonates and infants, *Wilmore et al.* (1969) used 4.3 mmol (100 mg) sodium/kg body weight/day when no fat was given. When fat is used as part of the intravenous diet, as little as 1–1.4 mmol (23–32 mg) sodium/kg/day may be sufficient for neonates and infants (*Grotte et al.,* 1976).

2. Potassium

The normal total body potassium is about 3,400 mmol, which is largely intracellular and freely exchangeable. The kidneys cannot reduce the loss of potassium as effectively as they can cope with sodium and it is therefore very important to correct for losses with an adequate daily intake. The basal daily requirement to prevent deficiency symptoms is estimated at between 0.7 and 0.9 mmol potassium/kg. Because of the high intracellular concentration of potassium, a much greater intake is required in anabolic states, and it has been recommended that potassium intake should be related to N intake at a level of 7 mmol potassium/g N (*Lee,* 1974).

For neonates and infants on intravenous nutrition with fat, 2 mmol potassium/kg/day are recommended by *Grotte* (1971). *Wilmore et al.* (1969) gave 3.9 mmol (0.15 g) potassium/kg/day as part of a total intravenous nutrition without fat (table XVI).

3. Calcium

The calcium requirements have been the subject of many investigations. A continuous calcium supply is necessary to form and maintain the skeleton. Furthermore, calcium must be supplied to preserve the extracellular ion concentration, the normal permeability of the cell membranes, and the optimal excitability of the nerve cells. The daily calcium losses total 320 mg or 8 mmol, equivalent to 0.11 mmol (4.5 mg)/kg in a person weighing 70 kg.

The negative calcium balance, which occurs owing to confinement to bed, cannot be compensated for by increasing the calcium supply. Nonetheless, for intravenous nutrition, a daily calcium supply of 0.11 mmol/kg is recommended in order to supply the amount of calcium which is normally absorbed from the ordinary food.

Larger quantities of calcium/kg body weight are required for neonates and infants than for adults. *Grotte* (1971) gave 0.5 mmol calcium/kg/day to neonates and infants. However, on the basis of oral balance studies, it has been estimated that the calcium requirement is 1 mmol/kg from birth until the age of 1 year (Recommended Dietary Allowances, 1968).

4. Magnesium

The magnesium content of the extracellular fluid is low compared with that of the intracellular fluid. A large portion of the magnesium reserve in the body is found in the bone. Magnesium is important for the function of a number of enzymes in carbohydrate metabolism, such as hexokinase and pyruvic acid oxidase in the brain, phosphorylation and other systems where thiamine pyrophosphate is a cofactor (*Davis,* 1964).

In magnesium deficiency, which may occur after a few weeks administration of magnesium-free infusion solutions, and more rapidly if there are gastrointestinal losses (*Paymaster,* 1975), neurological disturbances such as tetany are observed. The symptom usually depends on a concomitant deficiency in calcium and phosphate, but addition of magnesium is required for its cure as the condition is resistant to direct calcium therapy.

Large amounts of magnesium can cause general depression and even respiratory paralysis. However, this risk is only imminent in patients with renal insufficiency and treated with parenteral infusions, where elimination of magnesium is impaired.

An amount of 3 mmol, or 0.04 mmol/kg/24 h, should satisfy 'normal' adult requirements (*Wretlind,* 1972). To neonates and infants on intravenous nutrition, Børresen et al. (1970a) and Wilmore et al. (1969) gave 0.3 and 1 mmol magnesium/kg body weight/day, respectively. From calculations of the advisable intake for infants (Shils, 1972), a basal amount of 0.15 mmol magnesium/kg/day would be recommended to infants on intravenous feeding. However, because of the high intracellular concentration of magnesium, as with potassium, requirements are considerably higher in patients with net protein anabolism.

5. Iron

The administration of iron during intravenous nutrition is of particular importance, since the iron reserves of the body are usually very small due to the low iron content of the diet. Iron is an essential nutrient, and is present as a vital component of haemoglobin and myoglobin and of several enzymes such as cytochrome, catalase and peroxidases.

The iron requirement is considered to be 50 µmol (2.8 mg), for female patients of child-bearing age, and 18 µmol (1 mg) for post-menopausal women and adult men (FAO/WHO Expert Group, 1970). This corresponds to about 0.25–1.0 µmol/kg (Shils, 1972; Wretlind, 1972), although Jacobson and Wester (1977) have recently suggested that as little or 0.13 µmol/kg may be adequate. However, since most adult patients have depleted iron reserves, a daily supply of 50 µmol iron, or about 1 µmol/kg may be recommended during complete intravenous nutrition.

For infants, a daily intravenous supply of 2 µmol (0.11 mg) iron/kg body weight is apparently sufficient and corresponds to the oral intake recommended by Fomon (1967) for normal full-term infants, on the basis of a 10% absorption.

6. Manganese

Manganese is a component or an activator of enzymes, such as pyruvate decarboxylase, arginase, leucine aminopeptidase, alkaline phosphatase, and of the enzymes which participate in oxidative phosphorylation. In man, manganese deficiency has been associated with weight loss, dermatitis, changes in colour of hair and hypocholesterolaemia (Doisy, 1973). For adults, a daily intravenous manganese supply of 40 µmol (2.2 mg), or 0.6 µmol/kg body weight should more than cover the requirements, and corresponds to the quantity of manganese normally absorbed from a diet which contains about 73 µmol (4 mg) manganese (Sandstead, 1967). There are indications that this recommendation may be high, and provisionally we would suggest that 0.1 µmol/kg be given to meet basal requirements. However, in man there appears to be a very effective homeostatic mechanism which makes manganese one of the least toxic of the trace elements, and slight overdosage is therefore probably not harmful (Burch et al., 1975).

Ricour et al. (1975a) found that 0.3 µmol/kg/day was sufficient to maintain the blood concentration of manganese in infants. Wilmore et al. (1969) gave

0.7 μmol (0.04 mg) manganese/kg body weight/day to neonates and infants on total intravenous nutrition.

7. Zinc

Zinc is present in several enzymes, such as carbonic anhydrase, carboxy-peptidases, lactate dehydrogenase, and other dehydrogenases. A zinc deficiency is manifested as anaemia, splenomegaly, small body, hypogonadism, and geophagy (*Underwood*, 1971). A typical zinc deficiency dermatitis has been observed in patients receiving parenteral nutrition (*Kay and Tasman-Jones*, 1975).

Coats (1969) suggested that a daily basal zinc supply, during intravenous nutrition, of about 20 μmol (1.3 mg), or 0.3 μmol/kg would be adequate. However, *Jacobson and Wester* (1977) have found that about 0.7 μmol/kg was necessary to achieve balance in their patients with virtually basal requirements. Zinc losses are often increased in patients on parenteral nutrition, hence this upper figure is recommended for most patients. Adequate zinc supplements are particularly important during the anabolic phase of recovery (*Kay et al.*, 1976).

Wilmore et al. (1969) used 0.6 μmol (40 μg) zinc/kg body weight/day for neonates and infants on complete intravenous nutrition. The zinc absorbed from the intestine in neonates receiving human milk containing 60 μmol zinc/l (*Fomon*, 1967) is about 1 μmol zinc/kg body weight/day.

8. Copper

Copper is required for the formation of copper-containing proteins and enzymes in the body, such as ceruloplasmin, cytochrome *c*-oxidase, tyrosinase, and monoamine oxidase (*Frieden et al.*, 1965).

Copper deficiency causes anaemia, skeletal deformation, demyelination and degeneration of the nervous system, pigmentation defects, reproduction disturbances, and cardiovascular damage. Deficiency may also produce hypocupraemia, and impairment of iron metabolism (*Mills*, 1972). Although no specific symptoms or signs have been associated with copper deficiency in parenteral nutrition, blood levels of copper have been observed to fall steeply (*Hankins et al.*, 1976.

A quantity of 31 μmol (2 mg) copper in the daily oral diet seems to cover the requirements. If inadequate absorption is taken into consideration, about 1/6 of this amount (5 μmol or 0.3 mg) should be sufficient for intravenous nutrition. A daily copper supply of 5 μmol, or 0.07 μmol/kg, is thus recommended for the adult on complete intravenous feeding. *Shils* (1972) reports the same amount as sufficient to keep serum copper and ceruloplasmin at normal levels.

Wilmore et al. (1969) gave 0.34 μmol (20 μg) copper/kg body weight/day to infants on complete intravenous feeding. About 0.3 μmol copper/kg/day given

intravenously to infants would correspond to the estimated oral requirement of
0.7–2.1 μmol (42–135 μg)/kg/day (*Underwood*, 1971).

9. Chromium

Chromium has recently been shown to be of importance for glucose tolerance, and this ion is a cofactor with insulin for normal glucose utilisation in rats
and mice. A glucose load causes a rise in plasma chromium in man (*Glinnsman et
al.*, 1966). Much of this chromium may be lost in the urine (*Schroeder*, 1968),
with possible depletion of biologically active chromium stores (*Mertz*, 1970).
The chromium content of the tissues of adult man are small (less than 6 mg).
The very small absorption of dietary chromium (0.5–3%) make estimates of
parenteral requirements very difficult (*Reinhold*, 1975).

10. Selenium

Although no specific selenium deficiency state has been recognised in man,
selenium is necessary to prevent a type of liver necrosis in rats and also a
syndrome of muscular dystrophy in lambs and calves. Selenium is required as a
cofactor for glutathione peroxidase in erythrocytes. The dietary requirement of
selenium depends upon the vitamin E intake since both have anti-oxidant properties (*Reinhold*, 1975). *Jacobson and Wester* (1977) found that 0.006 μmol/kg
would maintain selenium balance during complete parenteral nutrition including
fat emulsion. *Ricour et al.* (1975a) have shown that an intake of 0.04 μmol/kg is
necessary to maintain the blood concentration of selenium during total parenteral nutrition in infants.

11. Molybdenum

Molybdenum (existing in xanthine oxidase and aldehyde oxidase) seems to
be an essential nutrient for man. There are recent indications that excess of
molybdenum may have an adverse effect on purine metabolism. Basic data are
not available to indicate a recommended supply during intravenous nutrition.
Jacobson and Wester (1977) have found that a positive balance of molybdenum
may be obtained in adult surgical patients when about 0.003 μmol/kg/day is
supplied intravenously.

12. Chlorine and Total Anion Requirement

Chlorine as chloride is required to maintain the electrolyte balance in the
body tissues. The total daily amount of mineral cations recommended in
table XVI is 2–2.6 mEq/kg for adults. The amount recommended to cover the
phosphate requirement corresponds to 0.24 mEq. In this calculation, 1 mmol
phosphorus has been taken as equal to 1.8 mEq phosphate (*Camien et al.*, 1969).
With the intake of 0.7 g amino acids/kg as recommended, the sulphur supply will

equal 0.5 mEq sulphate from 0.12 mmol cysteine and 0.13 mmol methionine (from an adequate amino acid mixture such as Vamin; table VIIIb). The total non-chloride anions will thus be 0.74 mEq. The difference between the indicated cations and anions is therefore 1.3–1.9 mEq/kg body weight. Consequently, a chloride supply of 1.3–1.9 mmol/kg is recommended for adults. Where there is a tendency towards hyperchloraemic acidosis, some of the chloride can be replaced with acetate, as either sodium or potassium salt. If the sulphur supply is increased or decreased, with the amino acids cysteine and methionine, the chloride supply should be reduced or increased proportionally.

The recommended basal amounts of cations to infants, according to table XVI, is 4.3–6.8 mEq/kg body weight. The anions from the amino acids will amount to about 1.8 mEq/kg. The recommended phosphorus intake (0.4–0.8 mmol/kg) may produce 0.72–1.44 mEq anions/kg. The non-chloride anions produced is therefore 2.5–3.2 mEq. Thus, the difference between cations and anions may be 1.8–4.3 mEq, which should be covered by a supply of chloride. The values recommended by *Shils* (1972) agree well with these figures. The chloride intake must be carefully controlled so as to prevent an iatrogenic acidosis or, more rarely, alkalosis.

13. Phosphorus

Phosphorus is just as necessary for bone and teeth as calcium. Within 7–10 days after starting total intravenous nutrition with solutions lacking phosphate, adult patients were found to be significantly hypophosphataemic (*Travis et al.,* 1971). The mechanism most sensitive to phosphate deficiency is the metabolism of glucose in the red blood cells to lactic acid. The sequence of enzymatic steps in this anaerobic process seems to be inhibited at the 3-carbon stage, as there is an accumulation of fructose diphosphate, glyceraldehyde-3-phosphate and dihydroxyacetone phosphate and a decrease in glucose-6-phosphate, fructose-6-phosphate, 3-phosphoglycerate, 2-phosphoglycerate, phosphoenolpyruvate, 2-3-diphosphoglycerate and ATP. The conclusion by *Travis et al.* (1971), that it is the glyceraldehyde-3-phosphate dehydrogenase that is inhibited by the lack of phosphate, appears well founded, especially as the latter is known to be a cofactor of the enzyme.

A low concentration of 2,3-diphosphoglycerate is one of the important effects of phosphate deficiency. As shown by *Benesch and Benesch* (1967), a reduced amount of 2,3-diphosphoglycerate causes an increased affinity of oxygen for haemoglobin, impairing the transport of the oxygen to the tissues, causing hyperventilation, paresthesia, and mental disturbances. The erythrocytes are damaged leading to increased formation of spherocytes. Even haemolytic anaemia has been reported (*Shils,* 1972).

The amount of phosphorus, by weight, supplied to an adult should correspond to the calcium intake (Recommended Dietary Allowances, 1968). The recommended daily phosphorus supply during intravenous nutrition is, therefore, 10 mmol, or 0.15 mmol/kg for adults. This must, however, be regarded only as a very basal level since phosphorus requirements are bound to be much higher when increased amounts of energy and N are supplied.

For infants, about 0.4–0.8 mmol phosphorus/kg/day should be given intravenously to correspond with the above recommended supply of calcium. These quantities would give the desired weight ratio of 1.5 between calcium and phosphorus (Recommended Dietary Allowances, 1968). In a more recent paper, Bórresen (1974) reported that balance experiments using these allowances of phosphorus have given negative values, and suggested that the daily intravenous dose of phosphorus should be 1.5 mmol/kg. Ricour et al. (1975b) did not observe any cases of hypophosphataemia in 63 consecutive infants receiving a regime including 100 kcal/kg and 1.25 mmol phosphorus/kg/24 h.

When 2 g fat/kg body weight, as 20 or 10% Intralipid is given, 0.15 and 0.30 mmol phosphorus/kg body weight, respectively, is provided by the phospholipids included in the emulsion.

14. Iodine

Iodine is an essential nutrient, and is required in synthesis of the thyroid hormones, thyroxine and triiodothyronine. The daily supply for an adult should be about 1 μmol (0.015 μmol/kg) or 100–140 μg (Recommended Dietary Allowances, 1968). The estimated requirement for infants is about 0.04 μmol/kg, or 5 μg iodine/kg.

15. Fluorine

Oral fluorine intake varies widely, depending mainly on the fluorine content of the drinking water. The daily intake is about 50–90 μmol (1–1.5 mg) for adults and 25–30 μmol (0.5–1 mg) for infants, about three quarters of which are absorbed. Fluorine is necessary for maximal resistance to caries, and assists in the maintenance of a normal skeleton. Fluorine does not seem, however, to be a nutrient essential to support life in man and animals. To correspond, in intravenous nutrition, to the ordinary intake, 0.7 μmol fluorine/kg for adults and 3 μmol/kg for neonates and infants are recommended.

16. Sulphur

Sulphur in the body amounts to about 175 g, and is mainly found in the sulphur-containing amino acids, methionine and cysteine, and in chondroitin sulphate and similar compounds in cartilage. There are no recommendations on the supply of sulphur. An adequate sulphur supply seems to be provided when amino acid mixtures, with adequate quantities of the sulphur-containing amino acids, are given. About 17 mmol, or 0.5 g sulphur are included in 50 g of an adequate amino acid mixture.

Table XVII. Tentatively recommended daily amounts of vitamins in basal metabolic state and in conditions of increased need. The recommendations for basal, moderate and high supply are explained in the text to table XVI

	Allowances/kg body weight/day to adult			Allowances/kg body weight/day to neonates and infants		
	basal amounts[1]	moderate amounts[2]	high supply	basal amounts[1]	moderate amounts	high supply
Thiamine, mg	0.02	0.04	0.3[3]	0.05	0.1	2[5]
Riboflavin, mg	0.03	0.06	0.3[3]	0.1	0.2	0.4[5]
Nicotinamide, mg	0.2	0.4	2[3]	1	2	4[5]
Pyridoxine, mg	0.03	0.06	0.4[3]	0.1	0.2	0.6[5]
Folic acid, µg	3	6	6–9[4]	20	40	50[5]
Cyanocobalamin, µg	0.03	0.06	0.06	0.2	0.4	5[6]
Pantothenic acid, mg	0.2	0.4	0.4	1	2	
Biotin, µg	5	10	10	30	60	
Ascorbic acid, mg	0.5	2	25[3]	3	6	20[5]
Retinol, µg	10 (33 IU)	10 (33 IU)	20[4] (67 IU)	100 (333 IU)	100 (333 IU)	150[5] (500 IU)
Ergocalciferol or cholecalciferol, µg	0.04 (2 IU)	0.04 (2 IU)	0.1[4] (4 IU)	2.5 (100 IU)	2.5 (100 IU)	2.5[8] (100 IU)
Phytylmenaquinone, µg	2	2	2	50	50	150[5]
Tocopherol, IU	0.5[7]	0.75	1	1	1–1.5	1.5

[1] Wretlind (1972). [2] American Medical Association (1975). [3] Giovanoni (1976). [4] Greene (1975). [5] Dudrick et al. (1976). [6] Heird et al. (1976). [7] Horwitt (1976). [8] Not more than 10 µg or 400 IU/day.

17. Other Trace Elements

Cobalt is present in vitamin B_{12}. In other forms, cobalt does not appear to be essential for man. *Wilmore et al.* (1969) gave 0.24 μmol cobalt/kg body weight/day to patients on intravenous nutrition.

Other elements, such as *aluminium, antimony, arsenic, barium, boron, cadmium, lead, nickel, rubidium, strontium, tin* and *vanadium* have not been proven to be essential for man, although a number of these are known to be essential for various other animal species.

18. Conclusions

Technical improvements have made it possible to provide total parenteral nutrition to patients for several weeks and even for several months. In these cases, it is important to consider adequately the mineral requirements. Many of these elements play important structural roles or act as cofactors for enzymes in a variety of metabolic pathways, making it imperative to supply them to all patients on long-term intravenous nutrition. Deficiency symptoms of the various minerals have been discussed and some allowances regarding intake proposed. Considerable differences exist in the methods used to meet the mineral requirements. Some of the trace elements are already present in adequate amounts in the other infusion solutions used, and hence no further specific supplement is necessary (*Hankins et al.*, 1976). A number of 'cocktails' have been proposed in order to meet the requirements for the other minerals (*Shils*, 1972; *Hull*, 1974; *Wretlind*, 1975). However, it is apparent from this brief survey that the nutritional and metabolic implications of minerals are multiple and complex and numerous practical aspects are still unanswered.

F. Vitamins

The tentatively recommended allowances of vitamins to patients on intravenous nutrition are summarised in table XVII. The background for the basal amounts are given in the following section. In situations where the patient is already malnourished or where he has increased requirements for energy and N, it seems reasonable to assume that he will also have considerably increased requirements for the various vitamins, especially the water-soluble vitamins. However, the amount of this increase has not been accurately determined in controlled studies. The various levels of supply of vitamins found in the recent literature are compiled in table XVII.

1. Water-Soluble Vitamins

a) Thiamine. Thiamine is of importance for carbohydrate metabolism as a part of the coenzyme for decarboxylation of α-keto acids and in the utilisation of the pentoses in the hexose monophosphate shunt. The absorption of thiamine from the intestine is complete and, therefore, the same allowances can be recommended as for oral intake (Recommended Dietary Allowances, 1968). The 'biological half-life' is 10–20 days (*Greene,* 1972), hence early supplementation of parenteral infusions is required.

b) Riboflavin. Riboflavin is an essential part of several flavoprotein enzymes operative in oxidative processes. The addition of riboflavin to solutions for intravenous nutrition is important as symptoms of deficiency may appear already after 7 days (*Greene,* 1972).

c) Niacin. Niacin is usually provided for parenteral infusion as nicotinamide which is a component of nicotinamide adenine dinucleotide (NAD) and nicotinamide adenine dinucleotidephosphate (NADP). These coenzymes assist in the transfer of hydrogen in fatty acid synthesis, glycolysis and tissue respiration. Nicotinamide can be formed from tryptophan and it is calculated that 60 mg tryptophan can yield 1 mg nicotinamide, or correspond to 1 mg nicotinamide equivalent. This route of supply is probably of little importance in parenteral nutrition since the amino acid solutions used have a fairly low tryptophan content (table VIIIa, b). The minimum requirement of nicotinamide equivalents to prevent pellagra is considered to be 4.4 mg/1,000 kcal (*Goldsmith,* 1965). The recommended content of nicotinamide equivalent in the daily diet is 6.6 mg/1,000 kcal (Recommended Dietary Allowances, 1968), or 16 mg/10 MJ. This recommendation can also be followed for intravenous nutrition.

d) Pyridoxine. Pyridoxine and other compounds (pyridoxal and pyridoxamine) with vitamin B_6 activity have an important influence on protein, carbohydrate and fat metabolism. A particularly important function of vitamin B_6 derivatives is to catalyze transamination reactions. Moreover, pyridoxine is involved in certain decarboxylation, transmethylation and phosphorylation reactions. A vitamin B_6 deficiency prevents, among other things, normal metabolism of tryptophan. Pyridoxine deficiency may become apparent in little more than 7 days, with electroencephalogram and electrocardiogram changes being evident in less than 3 weeks (*Greene,* 1972).

e) Folic acid. Folic acid and other compounds (e.g. folic acid glutamates) with folacin activity are transformed in the body into tetrahydrofolic acid, which is a coenzyme for reactions in which single carbon atoms are transferred, such as in the syntheses of purines and pyrimidines. Lack of folacin causes megaloblastic anaemia. The depots of folic acid are small (5–12 mg), and slight nutritional disturbances may lead to deficiency (FAO/WHO Expert Group, 1970).

f) Vitamin B_{12}. Vitamin B_{12} and its derivatives (e.g. cyanocobalamin, hydroxocobalamin) are necessary for all cells in the body. The requirements seem to be highest for the cells in the bone marrow, nervous system and intestinal tract. The most important function of this vitamin is to contribute to the synthesis of deoxyribonucleic acid (DNA). Daily injections of 0.5–1 μg of vitamin B_{12} have given a maximal haematological reaction in patients with

Addison's pernicious anaemia, and maintained normal haematological values in these patients (*Herbert*, 1968).

g) Pantothenic acid. As a component of coenzyme A, pantothenic acid is necessary and of vital importance in the metabolism of carbohydrates and fatty acids, and for the synthesis of sterols, steroid hormones, porphyrins and acetyl choline. The pantothenic acid requirement in man has not yet been determined. Symptoms of deficiency from the nervous and gastrointestinal systems appear in man only when a pantothenic acid-free diet has been given for at least 12 weeks (*Hodges et al.*, 1959). The ordinary daily diet is estimated to contain between 10 and 15 mg.

h) Biotin. Biotin, among other things, is essential for carboxylation processes, the urea cycle, synthesis of aspartate, decarboxylation and deamination. A deficiency in man can only be produced by inactivating biotin with avidin, a glycoprotein in raw egg white. Although the daily need in man has not yet been determined, a quantity of 0.3 mg has been considered adequate (Recommended Dietary Allowances, 1968). During intravenous nutrition, a daily amount of 5 µg/kg for adults and 30 µg/kg for infants is recommended.

i) Ascorbic acid. Ascorbic acid participates in many hydroxylation reactions, one of the most important being the hydroxylation of proline to hydroxyproline, which is necessary for the formation of collagen required in wound healing (*Peterkofsky and Udenfriend*, 1965). Ascorbic acid is also necessary for the normal metabolism of tyrosine (*Rogers and Gardner*, 1949).

The minimal quantity of ascorbic acid necessary to prevent the occurrence of scurvy is between 6.5 and 10 mg (*Baker et al.*, 1969). The mean turnover of ascorbic acid has been estimated at 21.5 mg/day (*Tolbert et al.*, 1967). It is remarkable that symptoms of deficiency do not appear before 90 days on vitamin C-free diet (FAO/WHO Expert Group, 1970).

j) Choline. Choline functions in the body as a source of labile methyl groups. It is an essential nutrient for many laboratory animals However, no symptoms resulting from choline deficiency have so far been observed in man. The ordinary oral diet contains 0.5–0.9 g of choline/day. In fat emulsions, choline-containing phospholipids are included, the amount of choline/100 g fat being 1 g in Intralipid 10%, and 0.5 g in Intralipid 20%.

2. Fat-Soluble Vitamins

a) Vitamin A. Vitamin A (retinol) is necessary for normal vision and for the maintenance of normal function of the epithelial cells. Different methods have been used to estimate man's requirement of vitamin A. A group of experts at FAO/WHO (1967) recommended a daily supply in food, of 2,500 IU of vitamin A activity, or 0.75 mg of retinol. If a person in good nutritional condition is put on a diet free of vitamin A, it will on average take 1.5 years before deficiency symptoms appear (*Greene*, 1972). In oral nutrition, the provitamin A carotenoid may be used, but for intravenous nutrition only retinol should be given to supply vitamin A activity.

b) Vitamin D. The need for an exogenous supply of compounds with vitamin D activity (cholecalciferol, ergocalciferol) has been thoroughly discussed. Cholecalciferol is converted to the hormone, 1,25-dihydroxycholecalciferol by the liver and kidneys. The vitamin is necessary for the maintenance of the calcium and phosphate homeostasis and for a normal bone structure (*Bordier et al.,* 1968). Since in persons receiving intravenous nutrition, the formation of vitamin D in the skin from 7-dehydrocholesterol under the influence of sunlight will be reduced, a daily supply of 2.5 µg (100 IU) of cholecalciferol or ergocalciferol, which corresponds to 0.04 µg/kg body weight, is recommended for adults. This quantity covers the minimal requirement in the adult (Recommended Dietary Allowances, 1968), and has also been shown to be adequate in treating osteomalacia (*Dent and Smith,* 1969). For neonates and infants on intravenous nutrition, the daily supply should be 2.5 µg of cholecalciferol or ergocalciferol/kg body weight and for prematures 5 µg/kg/day. It is of some interest that similar levels of intake of vitamin D are required in both oral and intravenous nutrition, despite the fact that one major function of vitamin D is to aid the absorption of calcium and phosphorus from the gastrointestinal tract (*Dudrick et al.,* 1976).

c) Vitamin K. To maintain normal levels of prothrombin and other coagulation factors (VII, IX, X and possibly V) a supply of compounds with vitamin K activity is required. The compound which is present in food is phytylmenaquinone (vitamin K_1). Vitamin K is formed in healthy adults by certain bacteria in the intestine, probably in a sufficient quantity to cover the requirements. In a patient treated with antibiotics, who was on intravenous nutrition without a supply of vitamin K, signs of vitamin K deficiency were observed after 9 days (*Berthoud et al.,* 1966). The daily requirement of vitamin K in oral and intravenous nutrition is not known exactly. *Crim and Calloway* (1970) estimated the desirable oral intake to be about 2 mg. In persons depleted of vitamin K by starvation and antibiotic treatment, it has been shown that the minimal requirement lies between 0.03 and 1.5 µg/phytylmenaquinone/kg body weight/day given intravenously, or 2–100 µg for adults (*Frick et al.,* 1967). A daily quantity of 2 µg phytylmenaquinone/kg body weight should thus be sufficient for an adult on intravenous nutrition, to prevent the appearance of deficiency symptoms. To neonates and infants, a daily quantity of 50 µg phytylmenaquinone/kg body weight should be included in the intravenous nutrition programme. This corresponds to the recommendation of 1 mg phytylmenaquinone given parenterally per week to infants (*Fomon,* 1967). Because menadione and its derivatives with vitamin K activity can cause toxic effects (haemolysis, enzyme inhibition), they should not be used for intravenous nutrition.

d) Vitamin E. Of the 8 naturally occurring forms of vitamin E, the one with the highest biological activity is d-α-tocopherol. One of the major functions of vitamin E is as an antioxidant, particularly to protect polyunsaturated fats from destructive oxidations, and thus preserve the structural integrity of all the membranes. Although the precise clinical effects of vitamin E have still to be fully established, mild vitamin E deficiency (5 IU/day) has been shown to be associated with increased breakdown of erythrocytes and therefore possibly also of other tissue cells (*Horwitt et al.,* 1963). A daily intake of 30 IU (30 mg *dl*-α-tocopherol acetate) has been considered adequate for adult men (Recommended Dietary Allowances, 1968). However, *Horwitt* (1976) has discussed the difficulties in making a firm recommendation regarding intake. When the polyunsaturated fat intake is high, increased amounts of vitamin E may be required. During the preparation of soybean triglycerides, vitamin E is also extracted from the soybeans. When the soybean triglyceride-egg yolk phospholipid emulsion Intralipid is used in parenteral nutrition, 500 ml of 20% Intralipid will provide about 100 mg of total tocopherol, of which about 10% is in the α form, and about 60% in the γ form. This would correspond to about 30–35 IU of vitamin E activity (*Hove and Harris,* 1947). This amount (or the proportionate increase due to a higher level of fat infusion) should meet the requirements of all patients receiving parenteral nutrition.

3. Conclusions

By definition, vitamins are of vital importance and must, accordingly, be included in the intravenous nutrition programme. Suitable preparations for addition to intravenous infusions have been described (*Dudrick et al.,* 1972b; *Wretlind,* 1975).

Even if the normal individual has stores for some of the vitamins sufficient for months, most patients where intravenous nutrition is indicated are in poor condition, with deficient vitamin stores. There is every reason to predict that patients after operation, trauma and burns have vitamin requirements far above normal, even if the data have not yet been determined. In these situations, amounts of vitamins much higher than those required for normal maintenance should therefore be used.

V. Osmolality and Acid-Base Balance during Intravenous Nutrition

A. Osmolality

Since the movement of water between the different fluid compartments of the body is largely determined by the concentration of dissolved particles, i.e. osmolality, it is important not to change the osmotic pressure of the blood too rapidly when an intravenous nutrition is given. In general, isotonic solutions should be used. Hypertonic solutions have to be given slowly according to the

Table XVIII. Osmolality of plasma and some infusion solutions

	mOsmol/kg water
Plasma	290
0.9% NaCl	285
10% Fat emulsion (Intralipid)[1]	280
20% Fat emulsion (Intralipid)[1]	330
5% Glucose	278
10% Glucose	523
20% Glucose	1,250
30% Glucose	2,100
50% Glucose	3,800
Enzymatic protein hydrolysate (3.3%) and glucose (5%) (Aminosol-glucose)[2]	555
Enzymatic casein hydrolysate (10%) (Aminosol 10%)[2]	926
Crystalline amino-acid mixture (7%) with fructose (10%) (Vamin)[2]	1,275

[1] Table X.
[2] Tables VIIIa, b.

technique devised by *Dudrick et al.* (1969). The osmolality of some infusion solutions is summarised in table XVIII. The osmolality of plasma has been given for comparison. It is evident that a large amount of fat emulsion can be given without any significant change in the osmotic pressure of the blood. *Bernhoff* (1970) has shown that the myocardial contractile force is very much reduced by rapid infusions of hypertonic carbohydrate solutions. Isoosmotic solutions, such as 5% glucose and 10 or 20% fat emulsion (Intralipid) caused no such effects.

It is important, however, to point out that although most solutions of glucose and of amino acids are very hypertonic and will therefore initially promote water retention, complete metabolism of these substances and of fat emulsion also leads to production of considerable amounts of free water. Catabolism of amino acids leads to the production of one molecule of urea for each two molecules of amino acid. To maintain osmotic balance, this free water and the urea must be excreted by the kidneys.

The osmolality of blood and tissues is very delicately regulated by the kidneys. Accordingly, the osmotic pressure of the body fluids is normally kept very constant at 290 ± 10 mosm/kg. As a result of the action of the kidneys in different conditions, dilute urine may only have one sixth of the osmolality of the blood plasma (50 mosm/kg), whereas after heavy sweating in disease, or where an inadequate fluid intake is provided, the urine may reach 5 times the

Table XIX. Acid formation from intravenous solutions

Infusion solution	Acid formation in excess of base		
	mEq/l	mEq/g amino acid	mEq/g fat
Intramin novum	121	1.8	–
Vamin	28	0.41	–
Aminosol 10%	26	0.26	–
Aminosol 3.3%	14	0.42	–
Intralipid 10%	27	–	0.27
Intralipid 20%	27	–	0.14

plasma osmolality (about 1,400 mosm/kg). Hormones play an important role in this osmotic regulation. The antidiuretic hormone from the hypophysis inhibits water excretion and lowers plasma osmolality in contrast to aldosterone which mainly retains sodium ions.

B. Acid-Base Balance

Enzymatic activity is very dependent on hydrogen ion concentration. Accordingly, the body attempts to maintain its hydrogen ion concentration virtually constant, with the plasma pH between 7.36 and 7.44.

In parenteral nutrition, patients not infrequently develop respiratory insufficiency, and quite often their renal function is impaired. In this way, they have difficulties to maintain normal acid-base balance. It is therefore important that all nutrients are given in an appropriate chemical form and in such proportions as avoid a dietary acidosis or alkalosis. In general, there is a greater risk of acidosis than of alkalosis. The reason for this is that the sulphur in methionine and cystine produces sulphate ions and the phosphorus in the phospholipids is metabolised to phosphate ions.

Heird et al. (1972a) observed acidosis in children after infusion of a mixture of synthetic *L*-amino acids. They concluded that the mixture contained an excess in the form of amino acids hydrochlorides, which after metabolism caused acidosis. They therefore recommended that arginine and lysine in amino acid mixtures should be given as the acetate salt rather than the chloride. The values in table XIX show that large amounts of acid may be formed from various infusion solutions.

In shock after trauma, major operations, burns and especially sepsis, metabolism is impaired, and metabolic acidosis frequently develops. *Clowes* (1971) points out that blood lactic acid concentration is a good indicator of the degree of shock and respiratory insufficiency. The danger of acidosis in renal insufficiency must be taken into account. Low pH and high blood lactate concentration may require bicarbonate therapy. *Beach et al.* (1974) regard a blood pH of 7.2 as the limit where bicarbonate therapy is mandatory. Caution must be employed in the use of bicarbonate with parenteral nutrition solutions since precipitation of mineral components may occur. This problem can be circumvented by using acetate instead of bicarbonate.

The risk of lactic acidosis after high doses of fructose is well known and needs special consideration in hypoxaemia, shock and uraemia. *Harries* (1972) recorded a plasma pH of 6.85 in 3 cases of lethal infantile lactic acidaemia.

Alkalosis, although relatively rare in patients receiving parenteral nutrition, may occur in hyperventilation, vomiting, potassium depletion and, iatrogenically, after excessive treatment of acidosis.

VI. Experimental Investigations in Animals on Intravenous Nutrition

This section will describe some experimental investigations of complete intravenous nutrition in animals. All infusion solutions to be used in parenteral nutrition should follow the general regulations for pharmaceuticals. The solutions must be tested and investigated in many ways to ensure that they are safe, and give the expected and optimal effects from a nutritional point of view. Animal studies are an integral part of this process of evaluation of preparations.

A. Experiments on Dogs

Rhode et al. (1949) infused hypertonic glucose and protein hydrolysate into dogs for from 4 to 20 weeks. Later on, *Dudrick et al.* (1967) performed a complete intravenous nutrition with all nutrients except fat in beagle puppies. The authors were not only successful in keeping them alive for several months, but they also managed to obtain normal growth and development. Soon afterwards, *Wilmore and Dudrick* (1969c) studied dogs after resection of large parts of the intestine. The dogs were fed post-operatively exclusively by vein. In this way, they reconstructed the clinical indication that was most important for surgeons.

Studies of complete intravenous alimentation including fat have been made in dogs during a 10-week experimental period (*Håkansson et al.,* 1967). To reduce the infused volumes, as much as 76% of the energy requirement was given as fat. The 78 kcal/kg/day were supplied in only 47 ml/kg water. All the dogs were alert and in good condition throughout the 10-week infusion period. Nothing abnormal was observed in the appearance or behaviour of the dogs at any time. On the average, the haemoglobin decreased to 88% of the initial values. The average N, sodium, potassium and calcium balances were positive.

In other experiments on dogs, body weight and positive N balance were maintained during periods up to 8 weeks (*Holm et al.,* 1975). No abnormalities were observed. Up to 88% of the energy was provided as fat. The N balance was not affected by a change in the non-protein intravenous energy supply from both fat (83 kcal%) and carbohydrate (17 kcal%) to carbohydrate alone.

One dog has been maintained on complete intravenous nutrition with 40% of the total energy supply from fat (Intralipid) during the whole period of gestation (*Holm et al.*, 1975). The dog was given the infusion through a catheter in the superior vena cava. The solutions given were a complete, crystalline amino acid mixture (Vamin), Intralipid and glucose together with vitamins and minerals. The mating occurred both on the day before and the day after the start of the complete intravenous nutrition. After 61 days, the dog gave birth to 6 puppies. One puppy was still-born, and 2 others died soon afterwards. The cause of death was later shown to have been vitamin K deficiency. The remaining 3 puppies stayed alive and did well. During the whole period of pregnancy there was a positive balance of N, sodium, potassium, calcium and magnesium. This study indicates that the fat emulsion does not produce any harmful effects on the fetus. It also proves that we have a reasonably good knowledge of the nutrients required for satisfactory nutrition during the period of rapid growth of the fetus.

B. Experiments on Rats

Daly et al. (1970) worked out a technique for feeding rats continuously by vein, and were able to show that protein deficiency led to increased loss of weight, to slow healing of anastomoses between intestines and to suture insufficiency. Later experiments by *Steiger et al.* (1972) confirmed the earlier results, and it was also found that protein deficiency caused intense fatty infiltration of the liver.

Wretlind and Roos (1975) have maintained rats for periods of 28 days on complete intravenous nutrition including fat, and have obtained satisfactory growth and development.

VII. Infusion Technique

The parenteral nutrition is performed by infusions through either a cannula in a peripheral vein or a central venous catheter in the caval vein. In most cases, the daily infusions are given continuously for 24 h. However, some authors prefer to give the infusions to adults only during a 12-hour period, either during the night or in day-time. The central venous catheter is filled with a heparin solution during the intermissions of the infusions (*K. Bergström et al.*, 1972). Alternatively, this 'rest' period may be used for infusion of amino acid solutions only, a rhythmic infusion system called cyclic hyperalimentation

(*Blackburn et al.*, 1975; *Maini et al.*, 1976). The cyclic feeding is supposed to maintain a good N utilisation and prevent dysfunction of the liver. The value of this infusion system has still to be fully assessed. In neonates and infants, the infusions have always been made continuously during 24 h/day.

A. In Adults
1. Infusion in Peripheral Vein

A complete parenteral nutrition is always given by the intravenous route. The most simple way to do this is to infuse into a peripheral vein using a small cannula.

Adequate nutrition can, as already mentioned, be given through a peripheral vein only when fat emulsion is used as the main energy source. The fat emulsion, the glucose and the amino acid solution are given from the different bottles by separate infusion sets which are connected at one or two Y junctions and then simultaneously infused through one catheter in the vein (fig. 5).

Deitel and Kainsky (1974) have performed infusion in peripheral veins for periods up to 78 days by a 'lipid system' which consists of 3 g fat/kg, 1.5 g glucose/kg and all other nutrients.

A cannula, for intravenous nutrition including fat in a peripheral, vein should be left at the same place in the vein no longer than 8–12 h/day to reduce the risk of mechanical irritation leading to thrombophlebitis. When the infusion schedule is restricted in this way to the day-time with nightly intermissions, this resembles the usual diet rhythm more than the continuous intravenous alimentation. This also gives the patient rest during the nights with diminished urinary flow. Such an infusion schedule delivered by gravity drip and based upon fat, Intralipid, as part of the energy source has been found to work excellently for many years.

Intravenous nutrition has often to be continued for several weeks and it will frequently be difficult to find a suitable vein on the arms, legs, hands, or feet. Under such conditions, a central catheter has to be inserted.

2. Infusion in Central Vein

The most commonly used central catheters are those introduced percutaneously into the subclavian vein. This technique was developed by *Dudrick et al.* (1969), primarily because fat emulsion was not available in the United States at that time. Deprived of the possibility of using fat emulsion, the American investigators had to cover the energy requirements with glucose and, to avoid overhydration, they were forced to use very concentrated solutions (25–50%) glucose. These solutions have between 6 and 13 times the osmotic pressure of blood plasma. In a peripheral vein, such a highly concentrated solution will soon damage the vascular wall, causing thrombosis and stopping the infusion. In a large vein, the blood flow dilutes the hypertonic solution sufficiently to avoid

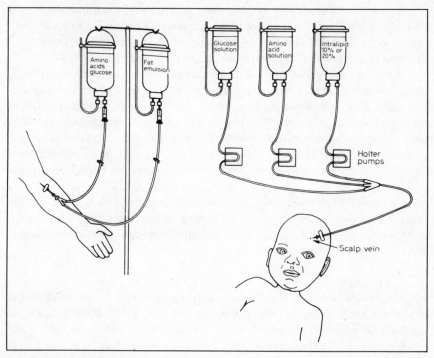

Fig. 5. Parenteral nutrition using a peripheral vein in adults (left) and neonates and infants. With adults, gravity drips are usually adequate whereas with infants constant infusion pumps are preferred.

thrombosis. As a rule, the catheter is exchanged, and a new one inserted on the other side every fourth week (*Dudrick et al.,* 1972b).

Solassol et al. (1973) have reported a special technique for the implantation of the catheter via a branch vein of the vena subclavia or epigastrica. *Holm and Wretlind* (1975) advocate the use of a cut down on a peripheral vein when inserting a central vein catheter. It is generally agreed that the catheter tip should lie just above the right atrium, its position being carefully controlled radiographically.

Most patients with a subclavian catheter receive all their treatment in hospital. However, *Broviac* (1972) reported that 4 patients with subclavian catheters were obtaining their nutrition at home. To make this possible, the catheters have to be fixed by a Dacron cuff near the vein, and passed through a subcutaneous tunnel down to a point just below and medial to the mammilla on the right side. There, the catheters are closed by a heparin lock. The low exit point excludes air embolism.

In the informative report by *Jeejeebhoy et al.* (1973), the authors describe a 34-year-old housewife who had received complete parenteral nutrition through central veins for a period of 9 months in hospital and 14 months at home (now for more than 6 years at home) without serious complications.

However, a large number of complications in the form of thrombosis, embolism and infections have been reported when using catheters in a central vein (sect. X) (*Dudrick and Ruberg,* 1971). The success of parenteral nutrition via central vein is primarily a question of technical skill, management and team work. There is no doubt that the widely different figures for local complications, particularly in patients with sepsis, depend on the capability of the different departments. *Holm and Wretlind* (1975) have recommended a special technique which should be applied in order, among other things, to avoid bacteriological complications. Their procedure includes tunneling the exterior end of the catheter subcutaneously, meticulous handling of the connection of the drip-aggregate, changing the three-way tap and drip-aggregate daily, and antiseptic dressing of the skin every second day.

B. In Infants

Intravenous nutrition is given to infants mainly with the same technique as is used for adults. When peripheral veins are sufficiently well-developed, and fat emulsion is available, peripheral infusion is preferable, particularly when the infusions are likely to be necessary only for a few weeks or less.

For neonates, the veins in the scalp are recommended, and fine guage needles rather than cannulae are used (fig. 5). *Børresen et al.* (1970b) treated 32 neonates by this route with complete intravenous nutrition for between 3 and 40 days after advanced surgery. The same vein could be used continuously for 3–4 days and reused after a further few days. *Coran* (1973) has reported results in close agreement with those of the last-mentioned authors. The injection site in the scalp must be inspected several times a day, as slight leakage causes infiltration, with the risk of necrosis and infection. *Coran* (1973) did not experience any infection or thrombophlebitis in 4 infants less than 3 months old and suffering from severe gastrointestinal diseases. The mean infusion time was 30 days. He stated with satisfaction that in no patient had the infusions to be continued by a central catheter.

In some situations, however, a subclavian catheter seems to be unavoidable at present. Central venous catheters have been used in hundreds of infants by *Dudrick et al.* (1972a). To help prevent infection, the catheter has to be conveyed under the skin to an opening in the scalp.

As with adults, it is clear that equally successful results can be obtained with both peripheral vein infusions or central vein infusions, provided the necessary degree of experience and expertise are available in establishment and maintenance of the infusion.

Table XX. Infusion solutions with additions for adults on intravenous nutrition. The given amounts are intended to cover a basal requirement. When greater amounts are demanded, 1.5–2 times the cited volumes should be given

---

Solution 1

(a) *Solution of crystalline amino acids (7%) and carbohydrates (10%)*
     (Vamin)[1]   1,000 ml
     with the addition of a
(b) *Solution of electrolytes ('Addam Electrolyte Solution)*[1]   10 ml
     containing 5 mmol Ca, 1.5 mmol Mg, 50 μmol Fe, 20 μmol Zn,
     40 μmol Mn, 5 μmol Cu, 50 μmol F, 1 μmol I and 13.3 mmol Cl

Solution 2

(a) *Fat emulsion (Intralipid 20%)*[2]   500 ml
     with the addition of an
(b) *Emulsion of fat soluble vitamins ('Vitalipid Emulsion for Adult')*[1]   10 ml
     containing 0.75 mg retinol, 3 μg cholecalciferol and 0.15 mg
     vitamin K_1

Solution 3

(a) *Glucose solution 10% for intravenous nutrition*   1,000 ml
     with the addition of a
(b) *Solution of lyophilyzed water soluble vitamins ('Soluvit')*[1]   10 ml
     containing 1.2 mg thiamine, 1.8 mg riboflavin, 10 mg nicotinamide,
     2 mg pyridoxine, 0.2 mg folic acid, 2 μg cyanocobalamin, 10 mg
     pantothenic acid, 0.3 mg biotin and 30 mg ascorbic acid, and
(c) *Potassium phosphate solution ('Addex-Kalium')*[3]   15 ml
     containing 30 mmol K, 6 mmol P and 21 mmol acetate

---

[1] Vitrum, Stockholm.
[2] Intralipid 10% may be used in an amount of 1,000 ml and with 500 ml of glucose solution 20%, if no increase of the total water volume is desired.
[3] Pharmacia, Uppsala.

VIII. Intravenous Nutrition in Adults: Indications, Dosages and Clinical Investigations

Suggested amounts of nutrients to be given to adult patients on intravenous nutrition are given in the tables XVI and XVII. Our guide for the practical performance of a complete intravenous nutrition with the solutions and injections used are given in the tables XX and XXI. The amounts supplied correspond to the basal requirement. In many cases the amounts of water, non-protein energy and amino acids have to be increased by 50 or 100%. There is also often an indication to increase the provision of electrolytes and vitamins above the basal daily allowance. Wherever possible, the provision of the various nutrients

Table XXI. Amounts of nutrients to adults on intravenous nutrition with the solutions given in table XX

Energy and nutrients	Amount/day			
	solution 1 1,010 ml	solution 2 510 ml	solution 3 1,025 ml	total
Water, litres	0.94	0.38	0.97	2.3
Energy, kcal	650	1,000	410	2,060
Amino acids, g	70	–	–	70
Glucose or fructose, g	100	12.5[1]	100	213
Fat, g	–	106[2]	–	106
Sodium, mmol	50	–	–	50
Potassium, mmol	20	–	30	50
Calcium, mmol	7.5	–	–	7.5
Magnesium, mmol	3.0	–	–	3.0
Iron, μmol	50	–	–	50
Zinc, μmol	20	–	–	20
Manganese, μmol	40	–	–	40
Copper, μmol	5	–	–	5
Chloride, mmol	68.3	–	–	68.3
Phosphorus, mmol	–	7.5	6	13.5
Fluoride, μmol	50	–	–	50
Iodide, μmol	1	–	–	1
Thiamine, mg	–	–	1.2	1.2
Riboflavin, mg	–	–	1.8	1.8
Nicotinamide, mg	–	–	10	10
Vitamin B_6, mg	–	–	2	2
Folic acid, mg	–	–	0.2	0.2
Vitamin B_{12}, μg	–	–	2	2
Pantothenic acid, mg	–	–	10	10
Biotin, mg	–	–	0.3	0.3
Ascorbic acid, mg	–	–	30	30
Vitamin A, mg	–	0.75	–	0.75
Vitamin D, μg	–	3	–	3
Vitamin K_1, mg	–	0.15	–	0.15
Tocopherol, mg	–	100	–	100

[1] Glycerol.
[2] 6 g phosphatides.

should be carefully monitored by estimation of losses and changes in blood concentrations.

As a general rule, parenteral nutrition is indicated when the patient cannot obtain an adequate intake of nutrients by the oral or enteral routes. Besides the obvious fact that a starving person must be given food, it has been shown that starving individuals are more sensitive to infections, they are more likely to have impaired wound healing and deficiency diseases are imminent even if no definite signs are observable. In many cases, the indication for intravenous nutrition is fairly clear-cut and the benefit of parenteral nutrition is well established. There are also, however, borderline cases where the value of intravenous nutrition is questionable and a controlled investigation is preferable to convictions based on 'practical experience'.

A. Pre-Operative Conditions
1. Poor Nutritional Status
The condition of severely malnourished patients, and those with decubitus ulcers in the pelvic region, must be improved before operation. Often parenteral nutrition is then unavoidable.

2. Peptic Ulceration
Chronic peptic ulcers are surrounded by scar tissue which causes deformation and strictures in the stomach and duodenum. Such patients invariably have a long history of digestive disease leading to undernutrition. This may make operation dangerous. Preliminary intravenous nutrition for a few weeks, as recommended by *Dudrick et al.* (1969) and *Dudrick and Rhoads* (1971), will improve the general condition of the patient, and reduce the risk of postoperative complications.

3. Neoplasia
Malignant disease is generally associated with anorexia and inanition. Parenteral nutrition is useful to improve the nutritional condition of the patient prior to operation. This holds both for cancer of the gastrointestinal tract as well as for other organs.

Parenteral nutrition is also valuable in reducing the stress of radiation treatment. Furthermore, *Dudrick and Rhoads* (1971) have pointed out that cytostatics (fluorouracil, etc.) can be used in more efficient doses if the patient's general condition is improved. In a series of 10 cachectic patients with metastatic adenocarcinoma of the colon treated simultaneously with 5-fluorouracil and intravenous hyperalimentation (*Souchon et al.,* 1975), the incidence of gastrointestinal side-effects was significantly less than in comparable patients fed by mouth. The tolerance to 5-fluorouracil was doubled and the response rate was greater in the group on intravenous nutrition.

Table XXII. Peak daily nitrogen loss after operations

Operation	Loss of N g/day	Operation	Loss of N g/day
Minor surgery	4	Peritonitis, bile fistula	18
Appendicectomy	5	Ruptured aneurysm	
Cholecystectomy	6	Sepsis, renal failure	22
Partial gastrectomy	8–13	Severe trauma, sepsis	27
Vagotomy and pyloroplasty	15	Ulcerative colitis, fever, laparatomy	35
Peritonitis, fistula	16		

After *Randall* (1970).

Chemotherapy of cancer is often followed by severe gastrointestinal side-effects. *Souchon et al.* (1975) also performed an experimental study in rats on the tolerance to the antimetabolite 5-fluorouracil with oral or intravenous nutrition. Twice as many of the intravenously fed rats survived and there was a lower incidence of side-effects compared with the group receiving food orally. Intravenous nutrition may therefore help to maintain nutritional status during courses of chemotherapy.

B. Immediate Post-Operative Period
1. Some General Aspects of the Metabolic Response to Trauma
Operation as well as all kinds of trauma lead to a number of metabolic changes, and intravenous alimentation is indicated in several of these instances.

More than 80 years ago, *Malcolm* (1893) observed that patients after trauma had fever even if there was no infection. *Dubois* (1924) discovered that the reason for the rise in temperature was an increased resting metabolism. The metabolic changes were extensively studied by *Cuthbertson* (1930, 1972). He pointed out that in the early period after trauma (ebb phase) there is a depression of metabolism, but later there is accelerated metabolism with increased oxygen consumption and increased N excretion (flow phase).

The mean N loss after different operations and infections is indicated in table XXII compiled by *Randall* (1970). The extreme loss of 35 g N/day corresponds to more than 200 g protein, or about 1,200 g lean tissue mass daily. Moreover, the period of N loss may be prolonged in major operations, trauma, and infection.

Quantitatively, the post-operative metabolic rise is in conformity with the severity of the operation. *Kinney et al.* (1970a) found an increase in energy consumption of 10% after uncomplicated elective operations, of 25% after multiple fractures, of 20–50% to be associated with peritonitis and of 50–125% after severe burns. They have shown that in such patients receiving inadequate nutritional support, the bulk of the additional energy is derived from the fat stores, the increased breakdown of tissue protein being required to provide amino acids for gluconeogenesis.

The requirement for energy causes increased lipolysis (*Tweedle and Johnston*, 1971), but the consumption is so great that serum glycerol and triglycerides are reduced by about 30% (*Schultis and Geser*, 1970). The concentration of FFA in serum is normal (*Drucker*, 1972). The low respiratory quotient (RQ = 0.75–0.85) is a proof that the fat is being oxidised.

Blackburn et al. (1976) point out that there is a much larger pool of protein in skeletal muscle (about 4.5 kg) than in the viscera (about 1.5 kg), and hence some part of the loss of muscular mass after trauma is used by the body to build up visceral protein which is of more immediate importance for the patient. This is in agreement with *Peaston*'s (1974) finding that trauma causes a greater loss of N and reduction in liver mass than does starvation. The losses of N and potassium in simple starvation are found to be relatively small (*Clark*, 1967).

Shock and trauma cause an insulin resistance which, during the first week, may lead to hyperglycaemia (*Stoner and Heath*, 1973). In some patients, this results in reduced glucose tolerance for several weeks, evidently as a result of persistent insulin insufficiency (*Allison et al.*, 1968). In the serum, the content of insulin is high (*Allison*, 1972), indicating that it is mainly the anabolic effect of the hormone, and not its release from the pancreas which is inhibited. The action of the various catabolic hormones is presumably dominant at this stage. Nonetheless, as pointed out earlier, in many cases glucose oxidation is increased within a few days of suffering a severe trauma (*Kinney et al.*, 1970a, b). This glucose intolerance may lead to acidosis (*Stoner and Heath*, 1973). *Cahill and Aoki* (1972) warn against giving glucose to support the catabolic phase when there is pronounced insulin resistance, as shown by hyperglycaemia and acidosis. This may cause a marked osmotic diuresis leading to dehydration and even death.

The glucose intolerance, however, is apparently only relative, since *Dudrick et al.* (1969) have been very successful in treating all kinds of severely ill patients with large amounts of glucose as the only calorie source ('hyperalimentation'). In some cases, insulin must be added to the infusion to ensure peripheral uptake of the glucose.

The object of parenteral nutrition after operations and trauma is to provide sufficient amounts of energy substrates, amino acids, and all other nutrients in an attempt to maintain the cell mass and function of all organs. Where this is not possible because of extreme hypercatabolism, the net loss of body N should at least be minimised. The overall adequacy of the infusion regime can be monitored fairly readily by measuring urine urea N or total urine N (*Rutten et al.*, 1975). An inadequate supply of certain nutrients for some time may not necessarily mean that signs of deficiency will occur. It seems reasonable to be the future trend, however, always to give complete intravenous nutrition. In this way, deficiency symptoms and nutritional complications will be prevented in patients where, for different reasons, parenteral nutrition has to be continued for long periods.

2. Parenteral Nutrition in the Immediate Post-Operative Period

Because in the uncomplicated case, resting metabolic rate is only increased by about 10% (*Kinney et al.*, 1970a), it is generally agreed that for the first few days following operation most patients do not require any nutritional support other than fluid, electrolytes and about 100 g glucose/day. However, when a return to normal oral feeding is not established after 4–5 days, complete intravenous nutrition should be commenced. Alternatively, in patients who are already malnourished prior to operation, or in whom surgical complications suggest that the post-operative period may be prolonged, parenteral nutrition should be started as early as possible once the patient's general condition is stable, and usually within 24 h of the operation.

Wadström and Wiklund (1964) found an average negative N balance of 12 g N/day in patients after cholecystectomy when they gave them 10 kcal/kg/day. Another group of their patients received 10 kcal and 0.1 g N/kg/day, resulting in a negative N balance of 4.8 g N/kg/day. A third group given 35 kcal and 0.1 g N/kg/day had a mean negative N balance of only 1.2 g N/day. More recent reports by *Deligné et al.* (1974) and *Bozzetti* (1976) have confirmed their conclusion that an increase in either N input or in energy substrates leads to an improved N balance in the post-operative period.

Nonetheless, as pointed out earlier, it is much more difficult to achieve positive N balance in the immediate post-operative period than in simple starvation (sect. IV.B.7.b). However, by using adequate amounts of energy substrates and amino acids, positive N balances can be obtained (*Dudrick et al.*, 1969; *Bozzetti*, 1976; *Bark et al.*, 1976).

C. During Complications after Surgery

1. Wound Rupture

Good nutrition is a prerequisite for wound healing. An adequate supply of energy, amino acids, zinc, and other nutrients are important. When some of these nutrients are lacking, oedema occurs in the wound edges as a local indication of poor healing. As early as 1930, *Harvey and Howes* observed that surgical wounds in the stomach of the rat healed in relation to the state of nutrition.

Bozzetti et al. (1975) have demonstrated the improved collagen synthesis in the wounds of patients receiving adequate parenteral nutrition. The decisive

importance of good nutrition for enteral sutures is emphasised by *Daly et al.*
(1972).

2. Fistulas

Fistulas represent a group of disease complications where intravenous nutrition has had some of its most beneficial results. This is true as a general statement, but is most evident in high enterocutaneous fistulas. These are rare
sequelae of enteral inflammations, but not too infrequently they appear as a
most feared complication after operations. Suture leakage may cause digestive
fluid either directly or through a localized peritonitis to penetrate the skin. The
fistulas often cause large losses of fluid, electrolytes and protein. The liquid
corrodes the skin around the opening of the fistula, sometimes over a large area.
The prognosis for high enteral fistulas used to be poor. *Welch and Edmunds*
(1962) found 50–58% mortality for gastric, duodenal or small intestinal fistulas.
MacFadyen et al. (1973) report a lower figure for the mortality – 40–65% –
after having compiled the results from 5 reports. The causes of death were
electrolyte imbalance (78%), malnutrition (61%) and sepsis (67%) (*Dressner et al.,*
1971). Death was frequently contributed to by the necessity to operate upon these
grossly malnourished patients in an attempt to restore patency of the bowel.

Apart from generally improving nutritional status, the use of intravenous
nutrition has a marked effect in reducing the volume of fistula fluid within a few
days, often by more than 50% (*MacFadyen et al.,* 1973).

In most cases, these effects lead to spontaneous closure of the fistulas, and
in others they can be successfully operated (*Børresen,* 1966). *MacFadyen et al.*
(1973) studied 78 gastrointestinal fistulas of which 55 (70.5%) closed spontaneously following parenteral nutrition. 17 of the remaining 23 were successfully
operated. Six fistulas did not close and the patients died. The combined parenteral and surgical therapy thus seems to have reduced the mortality to less than
8%. Some most remarkable cases have been described, such as that of *Dressner et
al.* (1971), where a patient with a perforated duodenal ulcer underwent partial
gastrectomy. Later, he developed 5 fistulas and lost 20 kg in weight in 50 days.
After 32 days of parenteral nutrition, all the fistulas closed and the patient
gained 4 kg.

The site of the fistula is related to the probability of healing. Oesophageal,
gastric, duodenal and colonic fistulas have a higher likelihood of healing than
jejunal or ileal fistulas (*Aguirre and Fischer,* 1976).

Pancreatic fistulas can also be healed by intravenous nutrition. The effect is
mainly the same as in the enterocutaneous fistulas. *Dudrick et al.* (1970b) found
a beneficial effect 1 or 2 days after the infusions were started. *Dudrick and
Rhoads* (1971) observed an 80% reduction in the secretion due to pancreatic
fistulas and, at the same time, body weight increased. These results were confirmed by *Sedgwick and Viglotti* (1971).

3. Ileus and Peritonitis

Peritonitis and prolonged paralytic ileus are clear indications for the use of parenteral nutrition. After abdominal operations, the stomach is usually paralysed for at least 2 days, but the intestine will often start its movements after 12 h (*Cuthbertson*, 1972). Where paralytic ileus is continued for several days, *Brøckner et al.* (1964), have suggested that intravenous nutrition may shorten the time till recovery.

D. In Head and Neck Surgery

Stell (1970) has performed pharyngolaryngectomy on 19 patients with cancer in these regions. He had to resect the pharynx and larynx and to transplant intestine or stomach to the operation site. To avoid post-operative catabolism, he gave 17 MJ (4,000 kcal)/day intravenously, according to the results obtained by *Peaston* (1968a). After 5 operational casualties and 2 patients with metastases, 12 patients in sequence survived; one of them succumbed later from an intercurrent disease. Their average stay in hospital was 4 weeks – the minimum was 12 days.

Copeland et al. (1975) have reported the beneficial effects of parenteral nutrition in 23 patients with major head and neck tumours treated either with radical surgery or radiotherapy. They point out that the magnitude of the surgical procedure which can be performed is often dictated by the probable healing capacity of the resultant wound.

Sudjian (1974) has shown that parenteral nutrition leads to a considerably better prognosis than tube feeds in patients following pharyngo-laryngectomy.

E. Renal Insufficiency

In acute and chronic renal insufficiency, metabolism is restricted to leaving only a small margin of relative safety. There may be increased catabolism and, concurrently, a retention of N. A restricted diet will exaggerate the catabolism, and an increase in the protein ingested will raise the urea content in the blood, with the impending risk of severe uraemia.

In this difficult situation, the clinician has to attempt to correct the metabolism, so that the accumulation of toxic products is inhibited, and the kidneys obtain time to recover, if any recovery is possible. The importance of N reduction in patients with renal disease has been recognised since the early 20th century (*Maddock et al.*, 1968). *v. Slyke et al.* (1930) could reduce the urea content in the blood by limiting the daily protein intake to 40 g. *Rose and Wixom* (1955) showed in young healthy men that urea could be used as a source of N for non-essential amino acid synthesis. This was confirmed in the clinic by *Giordano* (1963), and *Giovanetti and Maggiore* (1964), who demonstrated the value of a diet with essential amino acids and a sufficient amount of energy. In severe cases where there was uraemic gastroenteropathy with malaise, vomiting, diarrhoea, and lethargy, *Dudrick et al.* (1970a) succeeded in relieving the digestive symptoms by intravenous infusions of a special 'renal failure solution'. As little as 2–3 g of essential amino acids was found to be sufficient (*Wilmore and Dudrick,* 1969a). Blood urea N was lowered on the average to 50%, and hypocalcaemia, hyperphosphataemia, and acidosis were reversed. The metabolic analysis showed a positive N and potassium balance.

J. Bergström et al. (1972) have treated 20 uraemic patients with a special solution containing the 8 essential amino acids plus histidine (*Bergström et al.,* 1970). The patients could leave the hospital after a few weeks and continued oral amino acid therapy at home.

The precise value of parenteral nutrition in patients with renal failure still has to be fully assessed. In certain types of reversible acute renal failure, it seems already clear that the use of intravenous essential amino acid mixture leads to a higher incidence of survival and more rapid recovery (*Abel,* 1976). However, the mortality in surgical patients with acute renal failure remains high, the cause of death frequently being related to the primary disease, to sepsis or to haemorrhage. In chronic renal failure, the use of short periods of total parenteral nutrition to cope with particular situations may be beneficial. Essential amino acid mixtures are somewhat preferable to casein hydrolysates or complete amino acid mixtures since the rise in blood urea is more controlled and the frequency of dialysis can be reduced. As energy source, there is ample evidence that both fat emulsions and carbohydrate are well utilised in renal insufficiency.

F. Liver Disease

1. Complete Parenteral Nutrition in Liver Disease

Many patients with varying degrees of abnormality of liver function require parenteral nutrition. In general, a complete parenteral nutrition regime can be employed in all such patients provided the liver impairment is not so severe as to cause hepatic encephalopathy.

Liver insufficiency has been regarded with some reservation as a contraindication for intravenous fat emulsions (*Hallberg et al.,* 1967). The authors gave no definite reason for their opinion, but the accumulation of 'fat pigment' in the liver cells and the old experiences with cotton seed emulsions called for caution. As fat in the form of FFA, however, is the main calorie source for all cells – except central nervous system – it seems logical that fat would be metabolised even in cases of liver insufficiency.

Scattered observations from the literature support the view that patients with liver cirrhosis consume plasma fat more rapidly than normal persons (*Lawson,* 1967). *Zumtobel and Zehle* (1972) report that patients after operations utilised Intralipid without any adverse reactions even if they suffered from bile stasis, liver cirrhosis or chronic progressive hepatitis. *Michel et al.* (1976) investigated 16 patients with alcoholic cirrhosis who were given intravenous nutrition including Intralipid for 15 days. No effects on liver function could be related to the fat infusion.

2. Hepatic Encephalopathy

One particularly exciting area is the use of parenteral nutrition with carefully designed amino acid mixtures as a specific therapy in liver failure with encephalopathy. The generally held opinion that the cerebral symptoms were due to ammonia intoxication, resulting from the inability of the liver to convert ammonia to urea, was discredited by the demonstration of the lack of proportionality between plasma ammonia concentration and cerebral effects (*Fischer and James,* 1972). It has long been known that there are characteristic changes in the plasma amino acid pattern in severe liver disease (*Iber et al.,* 1957). The major changes are an elevation in the levels of phenylalanine, tyrosine, methionine, and to a lesser extent tryptophan, and a reduction in the levels of the branched-chain amino acids, leucine, isoleucine and valine. *Condon* (1971) indicated that these alterations may well be associated with the development of encephalopathy, since in dogs with porto-caval shunts fed varying types of diets, the shortest survival was observed in the group fed the highest amount of aromatic amino acids. *Fischer* and co-workers have pursued this theme, and have found that the parenteral use of a specially developed amino acid mixture containing low concentrations of aromatic amino acids but high concentrations of branched-chain amino acids together with large amounts of glucose led to normalisation of the plasma ratio of aromatic to branched-chain amino acids. This was associated with dramatic improvements in cerebral function of both dogs (*Fischer et al.,* 1975) and man (*Fischer,* 1976b). This effect may have been due to correction of the concentration of the various cerebral neurotransmitters (*Fischer et al.,* 1975). Whatever the precise mode of action of the therapy, it is clear that this is a major advance and further refinements of the solutions can be expected. However, since in most cases of hepatic failure the basic disease process is not reversible, the long-term prognosis of such severe liver disease is still very poor. An important general conclusion from the development of these special amino acid solutions for use in renal failure and hepatic failure is that no one amino acid preparation will produce optimal results in all patients, but rather the infused amino acids should be tailored to the requirements of the individual patient.

G. Burns

The N and energy consumption are exaggerated in burns partly because much body tissue is destroyed, but primarily because the skin itself is damaged over more or less extended areas. In severe burns, large quantities of water are lost. To evaporate this water, a large quantity of energy is needed — about 580 kcal/l of water. Naturally, the water losses are proportional to the burned surface and, in table XV (high supply), suggested figures for water and energy requirements are given. Moreover, the period of excessive water loss is prolonged in the severe cases (*Curreri,* 1972a). Careful grafting reduces evaporation and oxygen consumption (*Barr et al.,* 1969). This is confirmed by *Liljedahl* (1972) who also emphasises that good nutrition is necessary to enable the grafts to take.

The large losses of water, protein and calories must be covered by intake and, as a rule, natural eating and drinking are preferred. Most patients with relatively limited burns can be fed orally (*Moncrief,* 1970). As in all trauma, however, there is a degree of anorexia in burns (*Lamke et al.,* 1974) and besides this an early post-burn ileus may make intravenous nutrition preferable (*Curreri,* 1972a). Even with a good oral intake, *Curreri* (1972a) found the limit to be 3,500–4,000 kcal/day. Consequently, nutrients have to be given parenterally, and the large amounts of energy required make this very difficult without fat emulsions.

Fat emulsions are preferred to concentrated glucose solutions for several reasons: hypertonic solutions and large volumes increase mineral losses, glucose is often incompletely utilised owing to intolerance and the technique is more complicated. Fat emulsions, on the other hand, are completely utilised in burns; lipaemic plasma is cleared from 2 to 4 times more rapidly than normally (*Wilmore et al.,* 1973a).

Liljedahl (1972) analysed serum lipids in these patients and, although fat was infused, the contents of triglycerides and cholesterol were normal, and the free fatty acids, which, metabolically, constitute the most active part were reduced. These results confirm the rapid energy consumption. Only the phosphatides were found in higher concentrations, and had evidently not been eliminated as rapidly as the other components of the fat emulsion.

Wilmore et al. (1973a) found low phosphatidyl serine and a 70% reduction of all unsaturated fatty acids in the erythrocytes of several burnt patients treated without fat emulsions. These changes could be reversed by infusion of fat emulsion (Intralipid).

Mineral metabolism is also profoundly altered in burns. There is a retention of sodium as long as the N balance is negative, whereas excretion is increased when the balance becomes positive (*Allison,* 1972). As mentioned earlier (sect. IV.E.2), potassium metabolism is dependent on N balance (*Allison,* 1972; *Curreri,* 1972b). Immediately after a burn, approximately 6 mmol potassium/g N is excreted, and this is later reduced to 4 mmol/g (*Moncrief,* 1970).

The management of electrolyte intake is complicated by alterations in the distribution of sodium and potassium at least, across the cell membrane. Partial failure of the sodium pump mechanism leads to accumulation of sodium in the cells (*Allison,* 1974). This effect can be reversed when sufficient nutrition is given (*Curreri,* 1972b).

Suggested total intakes of all nutrients for patients with severe burns are given in table XV (high supply). Using a combination of early homograft cover, elevated temperature cubicles and combined high dose oral and intravenous nutrition for 4 months, *Lamke et al.* (1974) successfully treated 3 patients with 75–85% burns. None of the patients lost as much as 10% body weight. The results are a good illustration of the progress in treatment, since 40 years ago even a 30% burn was considered as lethal.

H. Unconsciousness and Paraplegia

In severe head trauma when the patient is unconscious intravenous nutrition may be required, although in many cases, tube feeds are satisfactory. *Alter* (1972) regards parenteral treatment as important in head injuries, but he also points out that clinical practice varies widely.

Dudrick et al. (1972b) found paraplegic debility with decubital ulcers to be an indication for parenteral nutrition when adequate enteral feeding could not be maintained.

I. Gastrointestinal Disorders

1. After Surgery on the Gastrointestinal Tract

If there is delay in return of gastrointestinal function after surgery, intravenous nutrition should be resorted to. Perhaps the single most clear-cut indication for parenteral nutrition is in the 'short-bowel syndrome' following massive intestinal resection. Several workers have successfully maintained patients with this syndrome in good nutritional condition for prolonged periods by intravenous nutrition (*Dudrick and Ruberg,* 1971; *Jacobson,* 1972a). In some cases, when insufficient bowel remains to adapt sufficiently for absorption of even elemental diets, long-term parenteral nutrition has been continued successfully at the patient's own home (*Scribner et al.,* 1970; *Jeejeebhoy et al.,* 1973; *Solassol and Joyeux,* 1976).

2. Enteritis

There are many enteral infections, with profuse diarrhoea of a subacute and chronic nature, which are indications for modern infusion therapy. *Johnston* (1971) regarded a gastrointestinal insufficiency of more than 5 days duration as an indication for parenteral nutrition.

3. Ulcerative Colitis and Regional Ileitis (Crohn's Disease)

In most patients with inflammatory bowel disease due either to ulcerative colitis or regional ileitis, there is a marked element of malnutrition. This has a rather complex aetiology, often being due to the combined effects of reduced food intake, malabsorption, and protein and blood loss into the bowel. The use of intravenous nutrition not only improves nutritional status, it also often markedly improves the general condition of the bowel (*Gimpel and Shilling,*

1970). In many cases, enterocutaneous fistulas close spontaneously, whereas in others the condition of the patient improves sufficiently to permit successful operation (*Hallberg et al.,* 1968; *Dudrick et al.,* 1971). Although the long-term progress of these conditions is still uncertain, the use of intravenous nutrition can generally at least lead to a temporary control of the diarrhoea and an abatement in the progress of the disease.

IX. Intravenous Nutrition in Infants: Indications, Dosages and Clinical Investigations

The problems, indications and benefits of intravenous nutrition in paediatrics are mainly the same as those for adults. There are, however, several questions regarding infants, and especially new-borns, which have to be considered. The growing child is not in 'equilibrium' in the same sense as the adult. The balance studies so often discussed for adult patients have to be modified for infants, and do not refer to a state where input and output are equal, but due attention should be paid to growth and development.

When the basal quantities of nutrients indicated in the tables XVI and XVII are supplied with an energy intake adjusted to maintain growth in infants, the nutritive requirements of a patient should be well covered. A practical guide for the amounts and types of infusion solutions as used by us is given in table XXIII. The amounts of energy and nutrients supplied in this way are shown in table XXIV. In many clinical conditions, there is an increased need of nutrients. After prolonged starvation, extra nutrients have to be added to replete the patient and to restore the body weight.

A. Premature Infants

Morphologically, the human brain is completely developed after the first 6—8 months of life. Accordingly, the appropriate nutrition of premature and new-born infants is of extreme importance (*Børresen,* 1974).

As a general rule, infants less than 1 year of age must have parenteral nutrition if inadequate oral intake prevents a normal increase in weight (*Børresen,* 1974). *Wilmore* (1972) also emphasises this and considers that a paediatric neurologist should assist in the care of these patients.

The rapid growth of the fetus during the last months of pregnancy shows that the accumulation of amino acids and the synthesis of protein proceed satisfactorily. Regarding energy, some doubt has been expressed as to whether the premature infant is able to metabolise fat, as there is very little fat delivered to it during its intrauterine life (*Børresen,* 1974). However, *Coran and Nesbakken* (1969) found that prematures utilised fat as well as adults do, at least if 50 IU heparin is added to the infusion/kg body weight. Moreover, *Grotte*

Table XXIII. Infusion solutions with additions for infants on intravenous nutrition

	Per kg per day
Solution 1	
(a) *Solution of crystalline amino acids (7%) and carbohydrates (10%) (Vamin)*[1] with the addition of a	30 ml
(b) *Solution of electrolytes*[1] containing 0.6 mmol Ca, 0.1 mmol Mg, 2 μmol Fe, 0.6 μmol Zn, 1 μmol Mn, 0.3 μmol Cu, 0.3 mmol P, 3 μmol F, 0.04 μmol I and 1.26 mmol Cl	4 ml
Solution 2	
(a) *Fat emulsion (Intralipid 20%)*[2] with the addition of an	20 ml
(b) *Emulsion of fat soluble vitamins*[1] containing 0.1 mg retinol, 2.5 μg cholecalciferol and 0.05 mg vitamin K_1	1 ml
Solution 3	
(a) *Glucose solution 10% for intravenous nutrition* with the addition of a	90 ml
(b) *Solution of lyophilyzed water soluble vitamins*[1] containing 0.12 mg thiamine, 0.18 mg riboflavin, 1 mg nicotinamide, 0.2 mg pyridoxine, 0.02 mg folic acid, 0.2 μg cyanocobalamin, 1 mg pantothenic acid, 0.03 mg biotin and 3 mg ascorbic acid and	0.5 ml
(c) *Sodium phosphate-potassium phosphate solution*[1] containing 0.1 mmol Na, 1.4 mmol K, 0.2 mmol Ca, 1.61 mmol Cl, 0.16 mmol P and 0.5 mmol lactate	1.4 ml

[1] Vitrum, Stockholm.
[2] Intralipid 10% in a volume of 40 ml may be used.

(1973), using a complete intravenous nutrition including fat, has succeeded in enabling premature infants to grow and develop quite normally. These results have been confirmed by *Cashore et al.* (1975) who obtained 19 survivors from 23 infants of birth weight less than 1,500 g. They infused up to 3.5 g fat/kg body weight/day, as part of the nutritional regime.

The question of the quantity of fat which may be given to new-borns has been widely discussed. *Gustafsson et al.* (1972) have carried out a series of investigations of fat tolerance in underweight new-borns. They found that prematures of normal weight for their age had a maximal capacity to eliminate fat from the blood plasma, of about 6 g fat/kg/day which is somewhat higher than for normal adults. 'Small-for-date' infants were not able to eliminate fat

Table XXIV. Amounts of nutrients to infants on intravenous nutrition with solutions given in table XXIII

Energy and nutrients	Amount/day			
	solution 1 34 ml/kg	solution 2 21 ml/kg	solution 3 92 ml/kg	total amount kg
Water, ml	32	16	85	134
Energy, kcal	20	40	36	96
Amino acids, g	2.1	–	–	2.1
Glucose, fructose or glycerol, g	3	0.5	9	12.5
Fat, g	–	4.2	–	4.2
Sodium, mmol	1.5	–	0.1	1.6
Potassium, mmol	0.6	–	1.4	2
Calcium, mmol	0.68	–	0.2	0.88
Magnesium, mmol	0.145	–	–	0.15
Iron, μmol	2	–	–	2
Zinc, μmol	0.6	–	–	0.6
Manganese, μmol	1	–	–	1
Copper, μmol	0.3	–	–	0.3
Chloride, mmol	2.91	–	1.61	4.52
Phosphorus, mmol	0.3	0.3	0.16	0.76
Fluoride, μmol	3	–	–	3
Iodide, μmol	0.04	–	–	0.04
Thiamine, mg	–	–	0.12	0.12
Riboflavin, mg	–	–	0.18	0.18
Nicotinamide, mg	–	–	1	1
Pyridoxine, mg	–	–	0.2	0.2
Folic acid, mg	–	–	0.02	0.02
Cyanocobalamin, μg	–	–	0.2	0.2
Pantothenic acid, mg	–	–	1	1
Biotin, mg	–	–	0.03	0.03
Ascorbic acid, mg	–	–	3	3
Vitamin A, mg	–	0.1	–	0.1
Vitamin D, μg	–	2.5	–	2.5
Vitamin K_1, mg	–	0.05	–	0.05
Tocopherol, mg	–	4	–	4

emulsion from the blood as well as the prematures. They develop a secondary hyperlipidaemia – in the chylomicrons and pre-β-fraction. In several cases, *Gustafsson et al.* (1972) found that an intravenous dose of heparin (50 IU/kg) definitely improved the capacity of small babies to eliminate emulsified fat. This finding supports the view that in infants, who have been undernourished during intrauterine life, there is some insufficiency in the function of the lipoprotein lipases. However, the metabolism of fat emulsion is clearly improved by use of heparin, making a high energy utilisation possible.

Scandinavian physicians evidently prefer infusions with fat emulsion for physiological and technical reasons. In the United States, however, where fat emulsions were not generally available, *Dudrick et al.* (1972a) have successfully treated very small prematures, weighing only 700–1,200 g, with fat-free solutions containing 20–25% glucose as energy source. *Benda and Babson* (1971) also used fat-free infusions and were able to avoid insertion of a catheter in the jugular or subclavian vein. For the first 1 or 2 days, they used injection into the umbilical vein and then continued with a scalp vein. The infants were given a solution containing 2% amino acids, 13% glucose, minerals and vitamins. Up to 78 kcal/kg could be infused without complications. After 5–22 days of parenteral nutrition, oral nutrition was slowly started, and increased by a few millilitres at every meal. In this way, the low-weight infants developed normally. *Børresen* (1974) suggests that a 20–30% excess of energy, and a 50% excess of N has to be given if the infusion solution does not contain linoleic acid.

It should, however, be pointed out that some workers believe that the use of parenteral nutrition may not necessarily improve the prognosis in very low birth weight premature infants and advocate controlled trials to assess the true value of parenteral nutrition in such cases (*Heird and Winters*, 1975).

B. Congenital Malformations in the Gastrointestinal Tract

Oesophageal and enteral atresia, gastroschisis and omphalocele are vital indications for operation and parenteral nutrition. *Dudrick and Rhoads* (1971) reported 70 such cases, and *Wilmore et al.* (1969) have treated 18 patients with a combined surgical and nutritional procedure, and were highly successful. *Das et al.* (1970) also reported good growth and, despite the high glucose infusions introduced by *Dudrick* and co-workers, blood glucose remained within normal limits. The authors report low values for FFA, glycerol and triglycerides in plasma, which is a physiological reaction when the energy is adequately covered by carbohydrate. *Coran* (1972) used infusions with fat emulsion after surgery for gastrointestinal malformations, and obtained good growth and positive N balance in more than 107 patients with life-threatening anomalies. He regards the possibility of giving total intravenous nutrition in a peripheral vein as 'a major finding'. Especially for these severely stressed and undernourished patients he considered it extremely valuable to be able to avoid a central catheter, hypertonic solutions and large volumes of fluid. Despite the inconveniences of hyperalimentation, however, *Dudrick and Rhoads* (1971) have been able to feed an infant parenterally for 18 months after resection of 97% of the bowel, on account of multiple atresias.

C. Pre-Operative Period

Infants with gastrointestinal and other diseases often suffer from malnutrition. Before operation is performed, nutritional status must be improved by oral or parenteral nutrition. The possibility of withstanding the stress of operation is much greater if the infants are in a good nutritional condition.

D. Post-Operative Period

The same reasons in favour of parenteral nutrition, as these presented earlier, apply to the period following operation in paediatric surgery. One striking difference between the results obtained with adults and infants is the relative ease with which a positive N balance can be achieved with infants in the immediate post-operative period.

Børresen et al. (1970a) have made balance determinations for infants in the post-operative state, and found 0.53 g N/kg/day, and 0.3–0.4 MJ (70–100 kcal)/kg/day necessary. In one study, 26 out of 32 patients who underwent major operations survived, whereas in a second study, all 14 patients treated survived. During the first 24 h after operation, only half of the computed amount of fluid, nutrients and electrolytes was given to prevent excess bronchial secretion and pulmonary oedema (*Børresen and Knutrud,* 1969). Even from the day of operation, however, their N balance was positive, and later the weight increased as for normal bottle-fed infants. *Grotte et al.* (1974) confirmed these results and emphasised the great capacity of infants after operation to eliminate fat emulsion from the blood, even more rapidly than after prolonged starvation.

Many investigations have been performed with 'hyperalimentation' (*Dudrick and Rhoads,* 1971) without fat after operations on the alimentary canal of infants. This method requires a central catheter which, in paediatric surgery, is preferably inserted in one of the jugular veins. 4 g/kg/day amino acids can be utilised, and of the high dose of 37 g glucose/kg/day, only 1% was lost in the urine (*Reiter,* 1971). *Heird et al.* (1972b) have compared parenteral nutrition with fat as the main energy source with the hyperalimentation technique. They found that not only was the technique including fat infusion less complicated, but also N appeared to be better utilised with this regime, since the same N balance could be obtained with a lower total N and energy input.

E. During Complications after Surgery

Filler et al. (1969) reported on 114 infants under 2 months of age with fistulas, peritoneal sepsis, post-operative enteral obstruction, ileus and chronic diarrhoea. These infants were treated with 'hyperalimentation' and all survived. This result was ascribed to the alimentation, since previously, similar patients had a high mortality. There was positive N balance and weight increase when no septicaemia intervened. Infants in a poor state of nutrition did not gain weight during the first few days, but later increase was normal.

A compilation of the use of intravenous nutrition including fat emulsions, during post-operative complications, has been made by *Grotte* (1973).

F. Liver and Renal Insufficiency

In liver insufficiency, well-balanced amino acid infusions are given to obtain a urea synthesis as low as possible. As a protection against ammonia intoxication 180 mg arginine/day is recommended (*Heird et al.*, 1972c).

In renal insufficiency, histidine and arginine are essential, and have to be included in the infusion solution. *Børresen* (1974) is of the opinion that infants with renal disease are not able to adequately synthesise the non-essential amino acids, and he therefore prefers to give a complete amino acid mixture where the relation between essential and non-essential amino acids is high. It is also important to supply adequate amounts of energy.

G. Inflammatory Bowel Disease

Hyman et al. (1971) described dramatically how 7 infants, less than 3 months old, died of persistent diarrhoea. The following 3 patients with a similar type of enteritis were given combined oral and intravenous nutrition, and survived. *Kjellmer* (1972) was able to save the lives of underweight infants with body weights less than 1,250 g who were suffering from diarrhoea. They were given an infusion with 0.3–0.4 MJ (75–100 kcal)/kg/day, half as fructose and the other half as fat emulsion. *Wei et al.* (1972) treated 23 patients aged from 3 months to 17 years with life-threatening gastroenteral diseases who were unable to eat. They were given complete intravenous nutrition with fat emulsion for 4 h, and all the water-soluble nutrients during the other 20 h; 0.5 MJ (122 kcal) in 145 ml of fluid/kg body weight was given daily.

H. Conclusions

In paediatrics, parenteral nutrition has undoubtedly reduced the mortality in many conditions. This is most evident in cases of congenital malformations and post-operative complications, such as fistulas, short bowel syndrome and severe enteral infections. It has been pointed out that the greater possibility of survival, shorter stay in hospital, fewer complications and, above all, improved brain development are all likely although it is difficult to obtain controlled statistical figures to confirm the clinical observations.

X. Adverse Reactions

A. Local Reactions
1. Peripheral Infusions

Hästbacka et al. (1966) have studied the frequency of thrombophlebitis in 1,048 infusions in hand veins. The incidence of thrombophlebitis was low when the infusion time was less than 8 h daily but rose steeply when the infusion was continued for more than 24 h. The infusion site, type of catheter, or pH of the solution appeared to be of less importance. In all, there were 25% of cases with thrombophlebitis (259 cases in 1,048 infusions). Such accidents as haematomas or perivenous infusion did not increase the frequency of thrombophlebitis.

Weiss and Nissan (1975) have also demonstrated a markedly reduced incidence of thrombophlebitis with infusion in a peripheral vein limited to a daily period of 12 h compared with continuous infusion.

2. Central Infusions

The insertion and maintenance of a central catheter is a much more risky operation than using a peripheral cannula and should only be used in specialised centres (*Savege*, 1973). The subclavian vein is preferred by the designers of the hyperalimentation technique (*Dudrick et al.*, 1969). However, in paediatrics the jugular vein is often recommended (*Johnston*, 1970). In the neighbourhood of these 2 veins, there are several structures which have been damaged during catheter insertion. A low percentage of pneumothorax has been reported, as well as subcutaneous haematomas, haemothorax, intrapleural infusion, puncture of the carotid artery, the thoracic duct and the brachial plexus. A list of these mechanical complications has been compiled by *Burri and Henkemeyer* (1972). Dislocation of catheters is not entirely uncommon and cardiac perforation has been reported in 9 cases (*Burri and Henkemeyer*, 1972).

The large diameter of the vessel and the speed of the bloodstream makes the occurrence of thrombosis much less frequent than in peripheral veins (*Savege*, 1973). *Dudrick* (1970) has had the same experience and reported no thrombosis or phlebitis in 200 patients or in dogs, which have a more rapid blood coagulation. *Wilmore and Dudrick* (1969b) had 2 infected catheters of 25 with infusion periods of 9–51 days. The infections with staphylococci generally come from the nose or *Candida albicans* from the mouth. If a catheter is moving slightly in the blood in a vein, and is covered by a fibrinous sheet, this is very easily infected, and *Broviac* (1972) found that the infection is mostly 'endogenous', starting from some infected focus elsewhere in the body. Mechanical damage to a vein can also lead to infection (*Fredrick and Guze*, 1971).

The occurrence of 'catheter embolism' is extremely infrequent. Nonetheless, *Burri et al.* (1971) have had 3 patients with this complication and have recorded 112 cases from the literature. The accident occurs when the puncture needle cuts off the tip of the catheter and the latter is carried away by the bloodstream. The mortality is 20–60% (*Burri et al.,* 1971). The authors have constructed a special needle which can be removed from the catheter. They also point out the importance of not inserting a catheter over a joint, such as the antecubital fossa.

B. General Reactions
1. Infection

Occurrence of septicaemia is the most feared complication of parenteral nutrition. In different studies, intravenous catheters have been believed to cause 9–43% of all nosocomial infections (*Frederick and Guze,* 1971). As local infection and phlebitis are very closely correlated with the infusion period, it is evident that the longer the infusion period, the higher is the risk of septicaemia. *Bentley and Lepper* (1968) report 0.7% of general infection if the catheter is *in situ* for less than 24 h and 2.5% for more than 48 h. In their review of 3,241 caval catheters, *Burri and Henkemeyer* (1972) demonstrated that the incidence of infection of the catheter tip rose to 25% when the catheter was continued for up to 14 days.

Since many patients receiving parenteral nutrition already have infections associated with abscesses, fistulas, wounds, etc., the aetiology of a particular septic episode may be very uncertain. In many cases, it can only be related to the catheter if the infection ceases on removal of the catheter, and culture of the catheter tip is positive. The infections are reduced, depending on the skill and care of the staff in the ward, the strictness of the regulations, and the discipline and standard of the hospital in general. There are great differences between various hospitals in these respects and, consequently, the figures for septicaemia vary within wide limits. At the extreme limits are *Curry and Quie* (1971) who found 27% of septic patients out of a total of 49, whereas *Owings et al.* (1972) reported no cases of septicaemia in 66 patients.

Obviously, the accountable microorganisms vary widely with the local conditions. *Fredrick and Guze* (1971) found mostly staphylococci, and several Gram-negative bacteria, while fungi were unusual. *McGovern* (1972), on the other hand, reported Candida to be most frequent, followed by *Staphylococcus aureus* and *Klebsiella aerobacter.* In a retrospective investigation, *Curry and Quie* (1971) found 22 patients with fungal septicaemia following total parenteral nutrition, 18 of whom died. In the investigation by *Dillon et al.* (1973), 83% of the infectious agents were either Gram-negatives or enterococci.

The large number of infections has led to general agreement that the insertion of a catheter should be regarded as a minor surgical operation with careful asepsis, including disinfection of the hands, the use of gloves, cap, gown, and protection of nose and mouth. The necessity of strict regulation is emphasised (*O'Neill,* 1972; *Holm and Wretlind,* 1975).

2. Metabolic Complications

An 'overloading syndrome' was earlier observed after 10–20 infusions with cottonseed oil emulsions (*Watkin and Steinfeld*, 1975). The symptoms were fever, jaundice, epigastric pain, bleeding in the gastrointestinal tract, anorexia and loss in weight. *Goulon et al.* (1974) have suggested an explanation of the overloading syndrome. They were able to discover antibodies to the soybean phosphatide, acting as emulsifier, in the serum of patients suffering from the overloading syndrome. When the antibodies complexed to the emulsifier, the emulsion was decomposed, and the liberated fat caused fat embolism which is associated with the above-mentioned symptoms.

a) Hyperglycaemia. Hyperglycaemia with glycosuria and osmotic diuresis is an important complication of hyperalimentation (*Wilmore et al.,* 1973a). In extreme cases, this may progress to hyperosmolar, hyperglycaemic, non-ketotic coma. If less than 0.5 g glucose/kg/h is infused in adult, however, all the infused glucose is utilised by most patients. Indeed, the intravenous glucose tolerance may be as high as 1.2 g/kg/h (*Dudrick et al.,* 1972a). As a certain part of the population has some degree of latent diabetes, and as there is frequently an element of insulin resistance in patients receiving parenteral nutrition, the advice given by the above-mentioned authors, always to control the glucose content in blood and urine, should be carefully followed.

b) Fluid overloading. Excess of fluid due to intravenous infusion can lead to circulatory overload, or pulmonary or cerebral oedema. This applies especially to patients with heart failure or renal insufficiency. The amount of fluid given ought not to exceed 135 ml/kg/day in new-borns, or 50 ml/kg/day in adults, except in cases of very high requirement.

c) Acidosis. A frequent, but usually not too severe complication is the development of a hyperchloraemic metabolic acidosis, due mainly to the use of large quantities of amino acid hydrochlorides. Monitoring of acid-base status and the use of potassium or sodium acetate permits ready control of the acidosis. Infusion of large amounts of fructose or ethanol may be associated with lactic acidosis.

d) Hyperammonaemia. Hyperammonaemia may occur, particularly in patients with hepatic disorders receiving casein hydrolysates. In such cases, crystalline amino acids are usually much better tolerated. Hyperammonaemia has also been associated with infusion of amino acid preparations containing inadequate amounts of arginine or excessive amounts of glycine.

e) Deficiency conditions. Deficiency conditions have to be carefully considered when parenteral nutrition is given for prolonged periods (more than 2 weeks). This is particularly so when the patient enters an anabolic phase, when the supply of intracellular ions may become limiting, e.g. potassium, magnesium zinc, and phosphate.

Attention must also be paid to minor minerals. Alteration in the blood levels of several of these has been reported (*Hankins et al.,* 1976). Close monitoring of the biochemical and haematological status of the patient should ensure that no deficiency state is allowed to become established.

XI. General Summary

After a historical review, some general physiological aspects on intravenous nutrition are presented. Then some technical methods and the requirements of patients are reported.

The metabolism of proteins, carbohydrates and fats is discussed, with special attention to the optimal balance, first, between the different amino acids and then between all the foodstuffs. The properties of the various preparations commonly available to supply these dietary components by the intravenous route are also described. During the course of the presentation, the connection between the basic science of nutrition and its clinical application is pointed out. The importance of *complete* parenteral nutrition with minerals and vitamins – including the essential fatty acids – is particularly emphasised.

The various clinical situations, where parenteral nutrition has been used are reviewed in detail, and some of the complications of such infusions are described.

In parenteral nutrition, special care must be taken in the design of the nutritional regime because the possible modulating influences of the intestine and of the liver are not able to take effect before the nutrients reach the systemic bloodstream. Like all new therapeutic methods, parenteral nutrition was extensively tested in animal models before it was considered safe enough for clinical trials. Each aspect of parenteral nutrition has been evaluated and improved over the years. Nutritionists have refined the methods for estimation of nutritional requirements in various conditions. Pharmacists have developed solutions capable of meeting these requirements. Surgeons have established new ways of inserting venous catheters, and nurses have developed procedures for maintaining these catheters patent and aseptic for long periods. The various clinical laboratories also have an important role in monitoring the effects of the infusions on the patient.

The clinical value of intravenous nutrition has been proven in adults and infants with a wide variety of conditions. There remain a number of areas where the precise value of parenteral nutrition, and the optimal components of this nutrition have still to be established. Although intravenous nutrition as a therapeutic procedure is still in its infancy, it is clearly growing rapidly and further advances can certainly be expected.

References

Abel, R.M.: Parenteral nutrition in the treatment of renal failure; in *Fischer* Total parenteral nutrition, pp. 143–170 (Little, Brown, Boston 1976).

Abitbol, C.L.; Fledman, D.B.; Ahmann, P., and Rudman, D.: Plasma amino acid patterns during supplemental intravenous nutrition of low birth weight infants. J. Pediat. *86:* 766–772 (1975).

Aguirre, A. and Fischer, J.E.: Intestinal fistulas; in *Fischer* Total parenteral nutrition, pp. 203–218 (Little, Brown, Boston 1976).

Ahnefeld, F.W.; Bässler, K.H.; Bauer, B.C., et al.: Suitability of non-glucose carbohydrates for parenteral nutrition. Eur. J. intens. Care Med. *1:* 105–113 (1975).

Allison, S.P.: Insulin and carbohydrates in parenteral feeding; in *Wilkinson* Parenteral nutrition, pp. 275–282 (Churchill, Livingstone, London 1972).

Allison, S.P.: Metabolic aspects of intensive care. Br. J. Hosp. Med. *11:* 862–872 (1974).

Allison, S.P.; Hinton, P., and Chamberlain, M.J.: Intravenous glucose, insulin, and free-fatty acid levels in burned patients. Lancet *ii:* 1113–1116 (1968).

Alter, H.: Wertung der dem Patienten zugeführten Infusionen besonders in Hinblick auf Fett- und Eiweisszufuhr. Hefte Unfallsheilkd. *III:* 258–260 (1972).

American Medical Association, Department of Foods and Nutrition: Guidelines for multi-vitamin preparations for parenteral use. Manuscript (AMA, Chicago 1975).

Artz, C.P.: Newer concepts of nutrition by the intravenous route. Ann. Surg. *149:* 841–851 (1959).

Ashby, M.M.; Heath, D.F., and Stoner, H.B.: A quantitative study of carbohydrate metabolism in the normal and injured rat. J. Physiol., Lond. *179:* 193–237 (1965).

Baker, E.M.; Hodges, R.E.; Hood, J.; Sauberlich, H.E., and March, S.C.: Metabolism of ascorbic-1-^{14}C acid in experimental human scurvy. Am. J. clin. Nutr. *22:* 549–558 (1969).

Bark, S.; Holm, I.; Håkansson, I., and Wretlind, A.: Nitrogen sparing effect of fat emulsion compared with glucose in the postoperative period. Acta chir. Scand. *142:* 423–427 (1976).

Barr, P.-O.; Birke, G.; Liljedahl, S.-O., and Plantin, L.-O.: Changes in BMR and evaporative water loss in the treatment of severe burns with warm dry air. Scand. J. Plast. reconstr. Surg. *3:* 30–38 (1969).

Bässler, K.H.: Physiological basis for the use of carbohydrates in parenteral nutrition; in *Meng and Law* Parenteral nutrition, pp. 96–108 (Thomas, Springfield 1970a).

Bässler, K.H.: Funktion der Kohlenhydrate in der Ernährung; in *Lang, Fekl and Berg* Balanced nutrition and therapy, pp. 31–36 (Thieme, Stuttgart 1970b).

Beach, F.X.M.; Wright, D.M., and Jones, E.S.: Acid-base balance and therapy; in *Lee* Parenteral nutrition in acute metabolic illness, pp. 177–195 (Academic Press, New York 1974).

Beeken, W.L.; Volwiler, W.; Goldsworthy, P.D.; Garby, L.E.; Reynolds, W.E.; Stogsdill, R., and Stemler, R.S.: Studies of I-131-albumin catabolism and distribution in normal young male adults. J. clin. Invest. *41:* 1312–1333 (1962).

Benda, G.I.M. and Babson, S.G.: Peripheral intravenous alimentation of the small premature infant. J. Pediat. *79:* 494–498 (1971).

Benesch, R. and Benesch, R.E.: The effect of organic phosphates from the human erythrocytes on the allosteric properties of hemoglobin. Biochem. biophys. Res. Commun. *26:* 162–167 (1967).

Bentley, D.W. and Lepper, M.H.: Septicemia related to indwelling venous catheter. J. Am. med. Ass. *206:* 1749–1752 (1968).

Bergström, J.; Bucht, H.; Fürst, P.; Hultman, E.; Josephson, B.; Norée, L.-O., and Vinnars, E.: Intravenous nutrition with amino acid solutions in patients with chronic uraemia. Acta med. scand. *191:* 359–367 (1972).

Bergström, J.; Fürst, P.; Josephson, B., and Norée, L.-O.: Improvement of nitrogen balance in a uremic patient by the addition of histidine to essential amino-acid solutions given intravenously. Life Sci. *9:* 787–794 (1970).

Bergström, J.; Fürst, P.; Norée, L.-O., and Vinnars, E.: Intracellular free amino acid concentration in human muscle tissue. J. appl. Physiol. *36:* 693–697 (1974).

Bergström, K.; Blomstrand, R., and Jacobson, S.: Long-term complete intravenous nutrition in man. Nutr. Metabol. *14:* suppl., pp. 118–149 (1972).

Bernhoff, A.: Negative inotropic effect of hyperosmotic infusion fluids. Opusc. Med. *15:* 162–164 (1970).

Berthoud, M.; Bouvier, C.A. et Krähenbühl, B.: Diagnostic différential d'une diathèse hémorragique aiguë: hypothrombinémie au cours d'une alimentation parentérale prolongée. Schweiz. med. Wschr. *96:* 1522–1524 (1966).

Beyriss, K.: Die parenterale Ernährung von Säuglingen und Kindern. Kinderärztl. Prax. *43:* 268–281 (1975).

Blackburn, G.L.; Flatt, J.P.; Clowes, G.H.A.; O'Donnell, T.F., and Hensle, T.E.: Protein sparing therapy during periods of starvation with sepsis or trauma. Ann. Surg. *177:* 588–594 (1973a).

Blackburn, G.L.; Flatt, J.P.; Clowes, G.H.A., and O'Donnell, T.E.: Peripheral intravenous feeding with isotonic amino acid solutions. Am. J. Surg. *125:* 447–454 (1973b).

Blackburn, G.L.; Flatt, J.P., and Hensle, T.W.: Peripheral amino acid infusions; in *Fischer* Total parenteral nutrition, pp. 363–394 (Little, Brown, Boston 1976).

Blackburn, G.L.; Maini, B.; Bistrian, B.R.; Flatt, J.P.; Page, G.; Sigman, D.; Stone, M., and Cochran, D.: Cyclic hyperalimentation – an optimal technique for preservation of visceral protein. Abstracts of Papers, Xth Int. Congr. Nutrition, Kyoto 1975, abstr., p. 237.

Bordier, P.; Matrajt, H.; Hioco, D.; Hepner, G.W.; Thompson, G.R., and Booth, C.: Subclinical vitamin D deficiency following gastric surgery. Histological evidence in bone. Lancet *i:* 437–440 (1968).

Børresen, H.C.: Total parenteral ernaering av en pasient med duodenalfistel. Tidsskr. norske Lægeforen. *86:* 1185–1192 (1966).

Børresen, H.C.: Balanced intravenous nutrition in pediatric surgery. Nutr. Metabol. Suppl. *14:* 114–117 (1972).

Børresen, H.C.: Clinical applications in paediatric surgery and paediatrics; in *Lee* Parenteral nutrition in acute metabolic illness, pp. 221–272 (Academic Press, New York 1974).

Børresen, H.C.; Coran, A.G., and Knutrud, O.: Metabolic results of parenteral feeding in neonatal surgery. Ann. Surg. *172:* 291–301 (1970a).

Børresen, H.C.; Coran, A.G., and Knutrud, O.: Parenteral feeding in the neonate; in *Berg* Advances in parenteral nutrition, pp. 93–96 (Thieme, Stuttgart 1970b).

Børresen, H.C. and Knutrud, O.: Parenteral feeding of neonates undergoing major surgery. Acta paediat. scand. *58:* 420–421 (1969).

Bozzetti, F.: Parenteral nutrition in surgical patient. Surgery Gynec. Obstet. *142:* 16–20 (1976).

Bozzetti, F.; Terno, G., and Longoni, C.: Parenteral hyperalimentation and wound healing. Surgery Gynec. Obstet. *141:* 712–714 (1975).

Brennan, M.F.; Fitzpatrick, G.F.; Cohen, K.H., and Moore, F.D.: Glycerol: major contributor to the short-term protein-sparing effect of fat emulsions in normal man. Ann. Surg. *182:* 386–394 (1975).

Brøckner, J.; Larsen, V., and Amris, C.J.: Early postoperative nutrition of surgical patients. Acta chir. scand. Suppl. *325:* 67–69 (1964).

Broviac, J.: Tubing, closures, infusion attachments, shunts; in AMA Symp. Total Parenteral Nutrition, Nashville 1972, pp. 160–170.

Broviac, J.W.; Riella, M.C., and Scribner, B.H.: The role of Intralipid in prolonged parenteral nutrition. I. As a calorie substitute for glucose. Am. J. clin. Nutr. *29:* 255–257 (1976).

Buchner, H. und Cesnik, H.: Erfahrungen über parenterale Fettzufuhr bei schwer Verletzten. Medsche Klin. *57:* 1482–1484 (1962).

Burch, R.E.; Hahn, H.K.J., and Sullivan, J.F.: Aspects of the roles of zinc, manganese and copper in human nutrition. Clin. Chem. *21:* 501–520 (1975).

Burr, G.O. and Burr, M.M.: On the nature and role of fatty acids essential in nutrition. J. biol. Chem. *86:* 587–621 (1930).

Burri, C. and Henkemeyer, H.C.: Review of the use of 3,241 caval catheters; in *Wilkinson* Parenteral nutrition, pp. 234–241 (Churchill, Livingstone, London 1972).

Burri, C.; Henkemeyer, H., and Pässler, H.H.: Katheterembolism. Schweiz. med. Wschr. *101:* 1575–1577 (1971).

Bünte, H.: Grundlagen der Schockbehandlung beim Ileus. Arch. klin. Chir. *308:* 187–189 (1964).

Cahill, G.F.: In discussion to H.N. Munro: Adaptation to mammalian protein metabolism to hyperalimentation; in *Cowan and Scheetz* Intravenous hyperalimentation, p. 52 (Lea & Febiger, Philadelphia 1972a).

Cahill, G.F. and Aoki, T.T.: The starvation state and requirements of the deficit economy; in *Cowan and Scheetz* Intravenous hyperalimentation, pp. 21–31 (Lea & Febiger, Philadelphia 1972).

Cailar, J. du; Crastes de Paulet, A.; Bomier, B.M.; Crastes de Paulet, P.; Kieulen, J. et Becker, H.: L'EB 51 (Trivémil) nutriment complet pour l'alimentation parentérale. Ann. Anésth. fr. spéc. *II:* 103–113 (1974).

Caldwell, M.D.; Meng, H.C., and Jonsson, H.T.: Essential fatty acids deficiency (EFAD) – now a human disease. Fed. Proc. *32:* 915 (1973).

Camien, M.N.; Simmons, D.H., and Gonick, H.C.: A critical reappraisal of 'acid-base' balance. Am. J. clin. Nutr. *22:* 786–793 (1969).

Carlson, L.A. and Hallberg, D.: Studies on the elimination of exogenous lipids from the blood stream. The kinetics of the elimination of a fat emulsion and of chylomicrons in the dog after single injection. Acta physiol. scand. *59:* 52–61 (1963).

Cashore, W.J.; Sedaghation, M.R., and Usher, R.H.: Nutritional supplements with intravenously administered lipid, protein hydrolysate and glucose in small premature infants. Pediatrics *56:* 8–16 (1975).

Chaptal, J.; Jean, R.; Crastes de Paulet, A.; Guillaumot, R.; Morel, G. et Maurin, A.: Apports énergétiques et azotés au cours du traitment des états de déshydration aiguë du nourrisson. Revue Prat., Paris *14:* 199–203 (1964).

Chaptal, J.; Jean, R.; Crastes du Paulet, A.; Pages, A.; Dossa, R.; Guillaumot, R.; Crastes de Paulet, P.; Robinet, A.; Morel, G. et Romeu, H.: Perfusions d'émulsions de lipides chez le nourrisson et l'enfant. Etude clinique, biologique et anatomique. Archs fr. Pédiat. *22:* 799–821 (1965).

Clark, R.G.: Surgical nutrition. Br. J. Surg. *54:* 455–459 (1967).

Clowes, G.H.A.: Oxygen transport and utilization in fulminating sepsis and septic shock; in *Hershey, del Guerco and McConn* Septic shock in man, pp. 85–105 (Little, Brown, Boston 1971).

Coats, D.A.: Complete parenteral nutrition. Lecture at Symp. Intravenous Therapy and Parenteral Nutrition, Melbourne 1969.

Collins, F.D.; Sinclair, A.J.; Royle, J.P.; Coats, D.A.; Maynard, A.T., and Leonard, R.F.: Plasma lipids in human linoleic acid deficiency. Nutr. Metabol. *13:* 150–167 (1971).

Condon, R.E.: Effect of dietary protein on symptom and survival in dogs with an Eck fistula. Am. J. Surg. *121:* 107–114 (1971).

Consolazio, C.F.; Johnson, R.E., and Pecora, L.J.: Physiological measurements of metabolic functions (McGraw-Hill, New York 1963).

Copeland, E.M.; MacFadyen, B.V.; MacComb, W.S.; Guillamondegui, O.; Jesse, R.H., and Dudrick, S.J.: Intravenous hyperalimentation in patients with head and neck cancer. Cancer *35:* 606–611 (1975).

Coran, A.G.: The intravenous use of fat for the total parenteral nutrition of the infant. Lipids *7:* 455–458 (1972).

Coran, A.G.: The long-term intravenous feeding of infants using peripheral veins. J. pediat. Surg. *8:* 801–807 (1973).

Coran, A.G.: Total intravenous feeding of infants and children without the use of a central venous catheter. Ann. Surg. *179:* 445–449 (1974).

Coran, A.G. and Nesbakken, R.: The metabolism of intravenously administered fat in adult and newborn dogs. Surgery, St Louis *66:* 922–928 (1969).

Crim, M.H. and Calloway, D.H.: A method for nutritional support of patients with severe renal failure. Nutr. Metabol. *12:* 111–120 (1970).

Cronberg, S. and Nilsson, I.-M.: Coagulation studies after administration of a fat emulsion, Intralipid. Thromb. Diath. haemorrh. *18:* 664–669 (1967).

Curreri, P.W.: Long-term supranormal caloric dietary program in extensively burned patients; in *Cowan and Scheetz* Intravenous hyperalimentation, pp. 135–144 (Lea & Febiger, Philadelphia 1972a).

Curreri, P.W.: Correction of elevated intracellular sodium concentration associated with major thermal trauma by the administration of combined enteral-intravenous dietary program; in *Cowan and Scheetz* Intravenous hyperalimentation, pp. 204–211 (Lea & Febiger, Philadelphia 1972b).

Curry, C.R. and Quie, P.G.: Fungal septicaemia in patients receiving parenteral hyperalimentation. New Engl. J. Med. *285:* 1221–1225 (1971).

Cuthbertson, D.P.: The disturbance of metabolism produced by bony and non-bony injury, with notes on certain abnormal conditions of bone. Biochem. J. *24:* 1244–1263 (1930).

Cuthbertson, D.P.: Protein requirements after injury – quality and quantity; in *Wilkinson* Parenteral nutrition, pp. 4–23 (Churchill, Livingstone, London 1972).

Daly, J.M.; Vars, H.M., and Dudrick, S.J.: Correlation of protein depletion with colonic anastomotic strength in rats. Surg. Forum *21:* 77–78 (1970).

Daly, J.M.; Vars, H.M., and Dudrick, S.J.: Effects of protein depletion on strength of colonic anastomoses. Surgery Gynec. Obstet. *134:* 15–21 (1972).

Das, J.B.; Filler, R.M.; Rubin, V.G., and Eraklis, A.J.: Intravenous dextrose-amino-acid feeding. The metabolic response in the surgical neonate. J. pediat. Surg. *5:* 127–135 (1970).

Davis, G.K.: Magnesium; in *Beaton and McHenry* Nutrition. A comprehensive treatise, vol. 1, pp. 463–481 (Academic Press, New York 1964).

Deitel, M. and Kaminsky, V.: Total nutrition by peripheral vein – the lipid system. Can. med. Ass. J. *111:* 152–154 (1974).

Deligné, P.; Prochiantz, E.; Bunodière, M.; Lauvergeat, J.; Brault, D.; Corcos, S.; Loygue, F. et Mailliard, G.: Bilanz azotés en période post-operatoire précoce de la chirurgie digestive. Influence des apports caloriques et protidiques (alimentation parentérale exclusive). Ann. Anésth. fr. spéc. *15:* 127–138 (1974).

Denis, J.B.: Lettre touchant deux expériences de la transfusions fites sur des hommes, pp. 12–13 (Cusson, Paris 1667).

Dent, C.E. and Smith, R.: Nutritional osteomalacia. Q. Jl Med. *38:* 195–209 (1969).

Dillon, J.D., jr.; Schaffner, W.; Way, C.W., van, II, and Meng, H.C.: Septicemia and total parenteral nutrition. Distinguishing catheter-related from other septic episodes. J. Am. med. Ass. *223:* 1341–1344 (1973).

Dölp, R.; Fekl, W. und Ahnefeld, F.W.: Freie Aminosäuren in Plasma in der posttraumatischen Phase. Infusionstherapie klin. Ernähr. *2:* 321–324 (1975).

Doisy, E.A., jr.: Micronutrient control on biosynthesis of clotting proteins and cholesterol; in *Hemphill* Proc. University of Missouri's 6th Annu. Conf. on Trace Elements in Environmental Health, pp. 193–199 (University of Missouri Press, Columbia 1973).

Dolif, D. und Jürgens, P.: Die Bedeutung der nichtessentiellen Aminosäuren bei der parenteralen Ernährung; in *Berg* Advances in parenteral nutrition. Symp. Int. Soc. Parenteral Nutrition, Prague, pp. 126–135 (Thieme, Stuttgart 1970).

Donahoe, J.F. and Powers, R.J.: Biochemical abnormalities with xylitol. New Engl. J. Med. *282:* 690–691 (1970).

Doolan, P.D.; Harper, H.A.; Hutchen, M.E., and Alpen, E.L.: The renal tubulae response to amino acid loading. J. clin. Invest. *35:* 888–895 (1965).

Dressner, S.A.; O'Grady, W.P., and Thorbjarnarson, B.: Parenteral hyperalimentation and multiple gastrointestinal fistulas. N.Y. St. J. Med. *71:* 665–669 (1971).

Drucker, W.R.: Carbohydrate metabolism. The traumatized versus normal states; in *Cowan and Scheetz* Intravenous hyperalimentation, pp. 55–66 (Lea & Febiger, Philadelphia 1972).

Dubois, E.F.: Basal metabolism in health and disease (Lea & Febiger, Philadelphia 1924).

Dudrick, S.J.: Growth, weight gain and positive nitrogen balance with long-term total parenteral nutrition; in *Meng and Law* Parenteral nutrition, p. 550 (Thomas, Springfield 1970).

Dudrick, S.J.; Copeland, E.M., and MacFadyen, B.V., jr.: Hyperalimentation in infants. Z. ErnährWiss. *15:* 9–25 (1976).

Dudrick, S.J.; Long, J.M.; Steiger, E., and Rhoads, J.E.: Technique of long-term parenteral nutrition and therapy; in *Lang, Fekl and Berg* Balanced nutrition and therapy, pp. 138–150 (Thieme, Stuttgart 1971).

Dudrick, S.J.; MacFadyen, B.V.; Buren, C.T. van; Ruberg, R.L., and Maynard, A.T.: Parenteral hyperalimentation. Metabolic problems and solutions. Ann. Surg. *176:* 259–264 (1972a).

Dudrick, S.J. and Rhoads, J.E.: New horizons for parenteral nutrition. J. Am. med. Ass. *215:* 939–949 (1971).

Dudrick, S.J. and Ruberg, R.L.: Principles and practise of parenteral nutrition. Gastroenterology *61:* 901–910 (1971).

Dudrick, S.J.; Steiger, E., and Long, J.M.: Renal failure in surgical patients. Treatment with intravenous essential amino acids and hypertonic glucose. Surgery, St Louis *68:* 180–186 (1970a).

Dudrick, S.J.; Steiger, E.; Long, J.M.; Ruberg, R.L.; Allen, T.R.; Vars, H.M., and Rhoads, J.E.: General principles and technique of administration in complete parenteral nutrition; in *Wilkinson* Parenteral nutrition, pp. 222–233 (Churchill, Livingstone, London 1972b).

Dudrick, S.J.; Vars, H.M., and Rhoads, J.E.: Growth of puppies receiving all nutritional requirements by vein; in Fortschritte der Parenteralen Ernährung, pp. 16–18 (Pallas, Lochham bei München 1967).

Dudrick, S.J.; Wilmore, D.W.; Steiger, E.; Machie, J.A., and Fitts, W.T.: Spontaneous closure of traumatic pancreatoduodenal fistulas with total intravenous nutrition. J. Trauma *10:* 542–553 (1970b).

Dudrick, S.J.; Wilmore, D.W.; Vars, H.M., and Rhoads, J.E.: Long-term total parenteral nutrition with growth, development, and positive nitrogen balance. Surgery, St Louis *64:* 134–142 (1968).

Dudrick, S.J.; Wilmore, D.W.; Vars, H.M., and Rhoads, J.E.: Can intravenous feeding as the sole means of nutrition support growth in the child and restore weight loss in an adult? An affirmative answer. Ann. Surg. *169:* 974–984 (1969).

Eckart, J.; Tempel, G. und Schürbrand, K.A.: Untersuchungen zur Utilisation parenteral verabfolgter Triglyceride nach Operationen und Traumen. Infusionstherapie klin. Ernähr. *2:* 138–145 (1973/74).

Elman, R.: Urinary output of nitrogen as influenced by i.v. injection of a mixture of amino acids. Proc. Soc. exp. Biol. Med. *37:* 610–613 (1937).

Elwyn, D.: The role of the liver in regulation of amino acids and protein metabolism; in *Munro* Mammalian protein metabolism, vol. 4, pp. 523–557 (Academic Press, New York 1970).

Engelberg, H.: Human endogenous lipemia clearing activity. Studies on lipolysis and effects of inhibitors. J. biol. Chem. *222:* 601–610 (1956).

FAO/WHO Expert Group: Protein requirements. FAO Nutr. Meet. Rep. Ser. No. 37 (1965).

FAO/WHO: Requirements of vitamin A, thiamine, riboflavine and niacin. FAO Nutr. Meet. Rep. Ser. No. 41, WHO Techn. Rep. Ser. No. 362 (1967).

FAO/WHO Expert Group: Requirements of ascorbic acid, vitamin D, vitamin B_{12}, folate, and iron. Techn. Rep. Ser. No. 452 (1970).

Feischl, P. und Hiotakis, K.: Vollständige parenterale Ernährung bei schwerem Tetanus. Medsche. Klin. *64:* 2296–2299 (1969).

Felig, P.: Amino acid metabolism in man. Annu. Rev. Biochem. *44:* 933–955 (1975).

Felig, P.: Intravenous nutrition: Fact and fancy. New Engl. J. Med. *294:* 1455–1456 (1976).

Filler, R.M.; Eraklis, A.J.; Rubin, V.G., and Das, J.B.: Long-term total parenteral nutrition in infants. New Engl. J. Med. *281:* 589–594 (1969).

Fischer, J.E.: Total parenteral nutrition (Little, Brown, Boston 1976a).

Fischer, J.E.: Personal commun. (1976b).

Fischer, J.E. and James, J.H.: Treatment of hepatic coma and hepatorenal syndrome: Mechanism of action of *L*-dopa and Aramine. Am. J. Surg. *123:* 222–230 (1972).

Fischer, J.E.; Yunovics, J.M.; Aguirre, A.; James, J.H.; Keane, J.M.; Wesdorp, R.I.C.; Yoshimura, N., and Westman, T.: The role of plasma amino acids in hepatic encephalopathy. Surgery, St Louis *78:* 276–290 (1975).

Fischer, P.: in *Fischer* Interkantonale Kontrollstelle. Ersatz-Zucker für die parenterale Ernährung. Grundsatz vom 19. März 1974. (1974.)

Fitzpatrick, G.F.; Meguid, M.M.; O'Connell, R.C.; O'Conner, N.E.; Ball, M.R., and Brennan, M.F.: Nitrogen sparing by carbohydrate in man. Intermittent or continuous enteral compared with continuous parenteral glucose. Surgery, St Louis *78:* 105–113 (1975).

Fleck, A.; Shepherd, J., and Munro, H.C.: Protein synthesis in rat liver. Influence of amino acids in diet on microsomes and polysomes. Science *150:* 628–629 (1965).

Foerster, H.; Hoos, I.; Boecker, S. und Michel, B.: Sind Maltoseinfusionen für die Infusionstherapie geeignet? Infusionstherapie klin. Ernähr. *2:* 385–392 (1975).

Foerster, H.; Meyer, E. und Ziege, M.: Erhöhung von Serumharnsäure und Serumbilirubin nach hochdosierten Infusionen von Sorbit, Xylit, und Fructose. Klin. Wschr. *48:* 878–879 (1970).

Fomon, S.J.: Infant nutrition (Saunders, Philadelphia 1967).

Forget, P.P.; Fernandes, J., and Haverkamp-Begemann, P.: Enhancement of fat elimination during intravenous feeding. Acta paediat. scand. *63:* 750–752 (1974).

Fox, I.H. and Kelley, W.N.: Studies of the mechanism of fructose-induced hyperuricemia in man. Metabolism *21:* 713–721 (1972).

Fraser, R. and Håkansson, I.: Personal commun. (1973).

Frazer, A.C.: Fat absorption and its disorders. Br. med. Bull. *14:* 212–220 (1958).

Fredrick, G.R. and Guze, L.B.: Infectious complications of intravenous polyethylene catheters. Calif. Med. *114:* 50–57 (1971).

Freuchen, I.: Intravenous fat emulsion. Nutr. Diet. *5:* 403–407 (1963).

Freuchen, I. und Østergaard, J.: Die parenterale Ernährung von Operationspatienten. Wien. med. Wschr. *117:* 49–51 (1967).

Freund, U.; Krausz, Y.; Levij, I.S., and Eliakim, M.: Iatrogenic lipidosis following prolonged intravenous hyperalimentation. Am. J. clin. Nutr. *28:* 1156–1160 (1975).

Frick, P.G.; Riedler, G., and Brögli, H.: Dose response and minimal daily requirement for vitamin K in man. J. appl. Physiol. *23:* 387–389 (1967).

Frieden, E.; Osaka, S., and Kobayashi, H.: Copper proteins and oxygen. Correlation between structure and function of the copper oxidases. J. gen. Physiol. Suppl. *49,* pp. 213–252 (1965).

Froesch, E.R.: Concluding remarks; in *Nikkila and Hattunen* Symp. Clinical and Metabolic Aspects of Fructose. Acta med. scand. Suppl. 542, pp. 239–244 (1972).

Froesch, E.R. and Ginsberg, J.L.: Fructose metabolism of adipose tissue. I. Comparison of fructose and glucose metabolism in epididymal adipose tissue of normal rats. J. biol. Chem. *237:* 3317–3324 (1962).

Froesch, E.R. and Keller, U.: Review of energy metabolism with particular reference to the metabolism of glucose, fructose, sorbitol and cylitol and of their therapeutic use in parenteral nutrition; in *Wilkinson* Parenteral nutrition, pp. 105–120 (Churchill, Livingstone, London 1972).

Galli, C.; White, H.B., jr., and Paoletti, R.: Brain lipid modifications induced by essential fatty acid deficiency in growing male and female rats. J. Neurochem. *17:* 347–355 (1970).

Gamble, J.L.: Chemical anatomy. Physiology and pathology of extracellular fluid (Harvard University Press, Cambridge 1958).

Garrow, J.S.: Recommended intakes of nutrients for the United Kingdom. Dept. of Health and Social Securities. Report on public health and medical subjects, No. 120, appendix 2 (HMSO, London 1969).

Gazzaniga, A.B.; Bartlett, R.H., and Shobe, J.B.: Nitrogen balance in patients receiving either fat or carbohydrate for total intravenous nutrition. Ann. Surg. *182:* 163–168 (1975).

Geiger, E.: Extra calorie function of dietary components in relation to protein utilization. Fed. Proc. Fed. Am. Socs exp. Biol. *10:* 670–675 (1951).

Geremy, F.; Nguyen-Huy-Dung et Deligné, P.: Etude par bilans azotés en periode catabolique post-operatoire d'une nouvelle solution de l'amino-acids (V 15 001 ou Vamin). Ann. Anésth. fr. *12:* 583–608 (1972).

Geyer, R.P.: Parenteral nutrition. Physiol. Rev. *40:* 150–186 (1960).

Geyer, R.P.; Chipman, J., and Stare, F.J.: Oxidation *in vivo* of emulsified radioactive trilaurin administered intravenously. J. biol. Chem. *176:* 1469–1470 (1948).

Ghadimi, H.: Total parenteral nutrition – Promises and premises (Wiley & Sons, Chichester 1975).

Gimpel, A. and Schilling, J.A.: Long-term hyperalimentation in ulcerative colitis. J. Okla. St. med. Ass. *63:* 371–375 (1970).

Giordano, C.: Use of exogenous and endogenous urea for protein synthesis in normal and uremic subjects. J. Lab. clin. Med. *62:* 231–246 (1963).

Giovanetti, S. and Maggiore, Q.: A low-nitrogen diet with proteins of high biological value for severe chronic uraemia. Lancet *i:* 1000–1003 (1964).

Giovanoni, R.: The manufacturing pharmacy solutions and incompatibilities; in *Fischer* Total parenteral nutrition, pp. 27–53 (Little, Brown, Boston 1976).

Glinnsman, W.H.; Feldman, F.J., and Mertz, W.: Plasma chromium after glucose administration. Science *152:* 1234–1245 (1966).

Goldsmith, G.A.: Niacin: antipellagra factor, hypocholesteremic agent. Model of nutrition research yesterday and today. J. Am. med. Ass. *194:* 167–173 (1965).

Goulon, M.; Barois, A.; Grosbuis, S. et Schortgen, G.: Embolie graisseuse après perfusions répétées d'émulsions lipidiques. Nouv. Presse méd. *3:* 13–18 (1974).

Greenberg, G.R.; Marliss, E.B.; Anderson, G.H.; Langer, B.; Spence, W.; Tovee, E.B., and Jeejeebhoy, K.N.: Protein sparing therapy in postoperative patients. New Engl. J. Med. *294:* 1411–1416 (1976).

Greene, H.C.; Hazlett, D., and Demaree, R.: Relationship between intralipid-induced hyperlipaemia and pulmonary function. Am. J. clin. Nutr. *29:* 127–135 (1976).

Greene, H.L.: Vitamins. AMA Symp. Total Parenteral Nutrition, Nashville 1972, pp. 78–91.

Greene, H.L.: Trace metals and vitamins; in *Winters and Hasselmeyer* Intravenous nutrition in high risk infants, pp. 273–282 (Wiley, New York 1975).

Grotte, G.: Nutrition parentérale du nourrisson; in *Nahas and Viars* Les solutés de substitution rééquilibration métabolique, pp. 509–515 (Librairie Arnette, Paris 1971).

Grotte, G.: Intravenous feeding in childhood. Postgrad. Med. *11:* 104–132 (1973).

Grotte, G.; Esscher, T.; Hambraeus, L., and Meurling, S.: Total parenteral nutrition in pediatric surgery. Proc. Meet., Vancouver, pp. 140–157 (Pharmacia, Quebec 1974).

Grotte, G.; Jacobson, S., and Wretlind, A.: Lipid and peripheral technique in parenteral nutrition; in *Fischer* Total parenteral nutrition, pp. 335–362 (Little, Brown, Boston 1976).

Gustafson, A.; Kjellmer, I.; Olegård, R., and Victorin, L.: Nutrition in low birth weight infants. Acta paediat. scand. *61:* 149–158 (1972).

Hadfield, J.: High caloric intravenous feeding in surgical patients. Clin. Med. *73:* 25–30 (1966).

Håkansson, I.: Experience in long-term studies on nine intravenous fat emulsions in dogs. Nutr. Diet. *10:* 54–76 (1968).

Håkansson, I.; Holm, I.; Obel, A.-L., and Wretlind, A.: Studies of complete intravenous alimentation in dogs; in Symp. Int. Soc. Parenteral Nutrition, Pallas, Lochham bei München 1967, pp. 11–15.

Hallberg, D.: Studies on the elimination of exogenous lipids from the blood stream. The kinetics for the elimination of a fat emulsion studied by single injection technique in man. Acta physiol. scand. *64:* 306–313 (1965a).

Hallberg, D.: Elimination of exogenous lipids from the blood stream. An experimental methodological and clinical study in dog and man. Acta physiol. scand. *65:* suppl. 254, pp. 1–23 (1965b).

Hallberg, D.; Holm, I.; Obel, A.-L.; Schuberth, O., and Wretlind, A.: Fat emulsion for complete intravenous nutrition. Postgrad. Med. *43:* 307–316 (1967).

Hallberg, D.; Schuberth, O., and Wretlind, A.: Experimental and clinical studies with fat emulsion for intravenous nutrition. Nutr. Diet. *8:* 245–281 (1966).

Hallberg, D.; Schuberth, O., and Wretlind, A.: Parenteral nutrition. Läkartidningen *65:* 4563–4574 (1968).

Hallberg, D. and Soda, M.: Effect of intravenous nutrients on hepatic blood flow; in

Romieu, Solassol, Joyeux and Astruc Comptes Rendus du Congr. Int. de Nutrition Parentérale, pp. 79–81 (Déhan, Montpellier 1976).

Halmagyi, M. und Isvang, H.H.: Auswahl der Kohlenhydrate zur intravenösen Anwendung in der intra- und postoperativen Phase; in *Lang, Frey and Halmagyi* Kohlenhydrate in der dringlichen Infusionstherapie, pp. 25–29 (Springer, Berlin 1968).

Hanc, I.; Klezkowska, H., and Rodkiewicz, B.: On intravenous infusions of fat emulsions in pediatric surgery. Ped. Polska *43:* 1355–1364 (1968).

Handler, P.; Kamin, H., and Harris, J.S.: Metabolism of parenterally administered amino acids; Glycine. J. biol. Chem. *179:* 283–301 (1949).

Hankins, D.A.; Riella, M.C.; Scribner, B.H., and Babb, A.L.: Whole blood trace element concentrations during total parenteral nutrition. Surgery, St Louis *79:* 674–677 (1976).

Hansen, A.E.: Serum lipid changes and therapeutic effects of oils in infant eczema. Proc. Soc. exp. Biol. Med. *31:* 160–161 (1933).

Harper, A.E.: Amino acid balance and food intake regulation; in *Meng and Law* Parenteral nutrition, pp. 181–217 (Thomas, Springfield 1970).

Harper, A.E. and Rogers, O.E.: Amino acid imbalance. Proc. Nutr. Soc. *24:* 173–190 (1965).

Harper, H.A.; Najarian, J.S., and Silen, W.: Effect of intravenously administered amino acids on blood ammonia. Proc. Soc. exp. Biol. Med. *92:* 558–560 (1956).

Harries, J.T.: Metabolic acidosis during intravenous feeding of infants; in *Wilkinson* Parenteral nutrition, pp. 266–274 (Churchill, Livingstone, London 1972).

Hartmann, G.: Fettemulsionen (Intralipid) in der inneren Medizin. Wien med. Wschr. *177:* 51–55 (1967).

Hartmann, G.: Praxis der parenteralen Ernährung. Z. prakt. Anästh. Wiederbel. *3:* 193 (1968).

Harvey, S.C. and Howes, E.L.: Effect of high protein diet on the velocity of the growth of fibroblasts in the healing wound. Ann. Surg. *91:* 641–650 (1930).

Hästbacka, J.; Tammisto, T.; Elfving, G., and Titenen, P.: Infusion thrombophlebitis. Acta scand. anaesth. *10:* 9–30 (1966).

Heird, W.C.; Dell, R.B.; Driscoll, J.M.; Grebin, B., and Winters, R.W.: Metabolic acidosis resulting from intravenous alimentation mixtures containing synthetic amino acids. New Engl. J. Med. *287:* 943–948 (1972a).

Heird, W.C.; Driscoll, J.M., jr.; Schullinger, J.N.; Grebin, B., and Winters, R.W.: Intravenous alimentation in pediatric patients. J. Pediat. *80:* 351–372 (1972b).

Heird, W.C.; MacMillan, R.W., and Winters, R.W.: Total parenteral nutrition in the pediatric patient; in *Fischer* Total parenteral nutrition, pp. 253–284 (Little, Brown, Boston 1976).

Heird, W.C.; Nicholson, J.F.; Driscoll, J.M.; Schullinger, J.N., and Winters, R.W.: Hyperammonemia resulting from intravenous alimentation using a mixture of synthetic *L*-amino acids. J. Pediat. *81:* 162–165 (1972c).

Heird, W.C. and Winters, R.W.: Total parenteral nutrition. The state of the art. J. Pediat. *82:* 2–16 (1975).

Henriques, V. und Andersen, A.C.: Über parenterale Ernährung durch intravenöse Injektion. Hoppe-Seyler's Z. physiol. Chem. *88:* 357–369 (1913).

Herbert, V.: Nutritional requirements for vitamin B_{12} and folic acid. Am. J. clin. Nutr. *21:* 743–752 (1968).

Hinton, P.; Allison, S.P.; Littlejohn, S., and Lloyd, J.: Insulin and glucose to reduce catabolic response to injury. Lancet *i:* 767–769 (1971).

Hodges, R.E.; Bean, W.B.; Ohlson, M.A., and Bleiler, R.: Human pantothenic acid deficiency produced by omega-methyl pantothenic acid. J. clin. Invest. *38:* 1421–1425 (1959).

Holm, I.; Håkansson, I.; Westling, K., and Wretlind, A.: Complete intravenous nutrition in dog (in Swedish). Opusc. Med. Suppl. *39:* 154–171 (1975).

Holm, I. and Wretlind, A.: Prophylaxis against infection and septicemia in parenteral nutrition via central intravenous catheter. Acta chir. scand. *141:* 173–181 (1975).

Holt, L.E., jr. and Snyderman, S.E.: The amino acid requirements of infants. J. Am. med. Ass. *175:* 100–103 (1961).

Hoover, H.C., jr.; Grant, J.P.; Gorschboth, C., and Ketchan, A.S.: Nitrogen-sparing intravenous fluids in postoperative patients. New Engl. J. Med. *293:* 172–175 (1975).

Hornstra, G.: Dietary fats and arterial thrombosis; in *Renaud and Nordøy* Dietary fats and thrombosis, pp. 21–52 (Karger, Basel 1974).

Horwitt, M.K.: Vitamin E: a reexamination. Am. J. clin. Nutr. *29:* 569–578 (1976).

Horwitt, M.K.; Century, B., and Zeman, A.A.: Erythrocyte survival time and reticulocyte levels after tocopherol depletion in man. Am. J. clin. Nutr. *12:* 99–106 (1963).

Hoshi, M.: Clinical application of sorbitol in patients with diabetes mellitus and in patients with liver disease. Med. J. Osaka Univ. *14:* 47–60 (1963).

Hove, E.L. and Harris, P.L.: Relative activity of the tocopherols in curing muscular dystrophy in rabbits. J. Nutr. *33:* 95–106 (1947).

Hull, R.L.: Use of trace elements in intravenous hyperalimentation solutions. Am. J. Hosp. Pharm. *31:* 759–761 (1974).

Hussein, M.L.; Young, V.R., and Scrimshaw, N.S.: Variations in endogenous nitrogen excretion in young men. Fed. Proc. Fed. Am. Socs exp. Biol. *27:* 485 (1968).

Huth, K.; Schoenborn, W. und Börner, J.: Zur Pathogenese der Unverträglichkeitserscheinungen bei parenteraler Fett-Zufuhr. Med. Ernähr. *8:* 146–148 (1967).

Hyman, C.J.; Reiter, J.; Rodnan, J., and Drash, A.L.: Parenteral and oral alimentation in the treatment·of the nonspecific protracted diarrheal syndrome of infancy. J. Pediat. *78:* 17–29 (1971).

Iber, F.L.; Rosen, H.; Levenson, S.M., and Chalmers, T.C.: The plasma amino acids in patients with liver failure. J. clin. Lab. Med. *50:* 417–425 (1957).

Jacobson, S.: Long-term parenteral nutrition following massive intestinal resection. Nutr. Metab. *14:* suppl., pp. 150–161 (1972a).

Jacobson, S.: Complete intravenous nutrition following massive intestinal resection. Int. Surg. *57:* 840–841 (1972b).

Jacobson, S.: The postoperative utilization of crystalline amino acid given intravenously in total parenteral nutrition. Proc. 19th Bienniel Congr. of the Int. College of Surgeons, Lima 1974.

Jacobson, S.: The postoperative electrolyte therapy after abdominal surgery. A balance study of sodium, potassium, calcium, magnesium and phosphorus in total parenteral nutrition (manuscript, 1976).

Jacobson, S.; Ericsson, J., and Obel, A.-L.: Histopathological and ultrastructural changes in the human liver during complete intravenous nutrition for seven months. Acta chir. scand. *137:* 335–349 (1971).

Jacobson, S. and Wester, P.-O.: Balance study of twenty trace elements during total parenteral nutrition in man. Br. J. Nutr. *37:* 107–126 (1977).

Jacobson, S. and Wretlind, A.: The use of fat emulsion for complete intravenous nutrition; in Body fluid replacement in the surgical patient, pp. 334–347 (Grune & Stratton, New York 1970).

Jeejeebhoy, K.N.: Personal commun. (1975).

Jeejeebhoy, K.N.; Anderson, G.H.; Nakhooda, A.F.; Greenberg, G.R.; Sanderson, I., and Marliss, E.B.: Metabolic studies in total parenteral nutrition with lipid in man. J. clin. Invest. *57:* 125–136 (1976).

Jeejeebhoy, K.N.; Zohrab, W.J.; Langer, B.; Philips, M.J.; Kuksis, A., and Anderson, G.H.: Total parenteral nutrition at home for 23 months without complications and with good rehabilitation. Gastroenterology *65:* 811–820 (1973).

Job, C. und Huber, O.: Zur Permeabilität menschlicher Erythrocyten für Glucose und Fructose. Naunyn-Schmiedebergs Arch. exp. Path. Pharmak. *241:* 53–54 (1961).

Johnson, D.G.: Total intravenous nutrition in newborn surgical patients. A three-year perspective. J. pediat. Surg. *5:* 601–605 (1970).

Johnson, J.D.; Albritten, W.L., and Sunshine, P.: Hyperammonaemia accompanying parenteral nutrition in newborn infants. J. Pediat. *81:* 154–161 (1972).

Johnston, I.D.A.: Parenteral feeding. Practitioner *206:* 103–110 (1971).

Johnston, I.D.A. and Spivey, J.: The use of long-term parenteral nutrients in alimentary failure; in *Berg* Advances in parenteral nutrition, pp. 82–87 (Thieme, Stuttgart 1970).

Johnston, I.D.A.; Tweedle, D., and Spivey, J.: Intravenous feeding after surgical operation; in *Wilkinson* Parenteral nutrition, pp. 189–197 (Churchill, Livingstone, London 1972).

Jones, E.; Robinson, J.S., and McConn, R.: Maintenance of metabolism during intensive patient care; in *Shuttleworth* Parenteral nutrition colloquium, pp. 32–34 (Lowe & Brydone, London 1964).

Jordal, K.: Proteinzufuhr in der parenteralen Ernährung in der Chirurgie. Int. Z. Vitam-Forsch. *35:* 26–38 (1965).

Jürgens, P. und Dolif, D.: Die Bedeutung nichtessentieller Aminosäuren für den Stickstoffhaushalt des Menschen unter parenteraler Ernährung. Klin. Wschr. *46:* 131–143 (1968).

Jürgens, P. and Dolif, D.: Experimental results of parenteral nutrition with amino acids; in *Wilkinson* Parenteral nutrition, pp. 77–92 (Churchill, Livingstone, London 1972).

Kaplan, M.W.; Mares, A.; Quintana, P.; Strauss, J.; Huxtable, R.F.; Brennan, P., and Hays, D.M.: High caloric glucose-nitrogen infusions. Postoperative management of neonatal infants. Archs Surg., Chicago *99:* 567–571 (1969).

Kapp, J.P.; Duckert, F., and Hartmann, G.: Platelet adhesiveness and serum lipids during and after Intralipid® infusions. Nutr. Metabol. *13:* 92–99 (1971).

Kausch, W.: Über intravenöse und subcutane Ernährung mit Traubenzucker. Dt. med. Wschr. *37:* 8–9 (1911).

Kay, R.G. and Tasman-Jones, C.: Zinc deficiency and intravenous feeding. Lancet *ii:* 605–606 (1975).

Kay, R.G.; Tasman-Jones, C.; Pybus, J.; Whiting, R., and Black, H.: A syndrome of acute zinc deficiency during total parenteral alimentation in man. Ann. Surg. *183:* 337–340 (1976).

Keller, U. und Froesch, E.R.: Vergleichende Untersuchungen über den Stoffwechsel von Xylit, Sorbit und Fruktose beim Menschen. Schweiz. med. Wschr. *102:* 1017–1022 (1972).

Kessler, E.: Hyperalimentation in the management of gastrointestinal cutaneous fistulae. S. Afr. J. Surg. *12:* 101–105 (1974).

Kinney, J.M.: Energy requirements for parenteral nutrition; in *Fischer* Total parenteral nutrition, pp. 135–142 (Little, Brown, Boston 1976).

Kinney, J.M.; Duke, J.H.; Long, C.L., and Gump, F.E.: Carbohydrate and nitrogen metabolism after injury. J. clin. Path. *23:* 65–72 (1970a).

Kinney, J.M.; Long, C.L., and Duke, J.H.: Carbohydrate and nitrogen metabolism after injury; in *Porter and Knight* Energy metabolism in trauma. Ciba Found. Symp., pp. 103–123 (1970b).

Kjellmer, I.: Intravenös tillförsel av kalorier till nyfödda och späda barn. Kätiölehti *77:* 263–265 (1972).

Knauff, H.G.: Amino acid solutions; in *Meng and Law* Parenteral nutrition, pp. 251–271 (Thomas, Springfield 1970).

Kofranyi, E.; Jekat, F.; Brand, E.; Hackenberg, K. und Hess, B.: Die Frage der Essentialität von Arginin und Histidin. Hoppe-Seyler's Z. physiol. Chem. *350:* 1401–1404 (1969).

Kohri, H.; Muto, Y., and Hosoya, N.: Metabolism of circulating maltose in guinea pig. J. Jap. Soc. Food Nutr. *25:* 641–646 (1972).

Krediet, R.T. and De Gens, A.: Parenchymateuze icterus en thrombocytopenie na langdurige behandling met een intraveneuze vetemulsie (Intralipid). Ned. Tijdschr. Geneesk. *119:* 1766 (1975).

Lamke, L.-O.; Liljedahl, S.-O. et Wretlind, A.: Aspects nutritionnels et cliniques de la nutrition intra-veineuse chez les brulés. Ann. Anésth. fr. *15:* spéc. II. pp. 27–35 (1974).

Lang, K.: Physiology and metabolism of amino acids; in *Meng and Law* Parenteral nutrition, pp. 160–180 (Thomas, Springfield 1972).

Lawson, L.J.: Parenteral nutrition in surgery. Br. J. Surg. *52:* 795–800 (1965).

Lawson, L.T.: Parenteral nutrition in surgery. Hosp. Med. *July:* 899 (1967).

Lee, H.A.: Parenteral nutrition in acute metabolic illness (Academic Press, New York 1974).

Lee, H.A. and Shortle, W.P.J.: Parenteral nutrition. King's Coll. Hosp. Gaz. *44:* 192–204 (1965).

Lemperle, G.; Reichelt, M., and Denk, S.: The evaluation of phagocytic activity in men by means of a lipid-clearing test. Abstr. 6th Int. Meet. of the Reticuloendothelial Soc., Mannheim 1970, p. 83.

Levin, G.; Roos, K.-A., and Kihlberg, R.: A comparison of enteral and parenteral nutrition in rats. Abstr. 10th Int. Congr. Nutrition, Kyoto 1975, p. 239.

Lidström, F. and Wretlind, A.: The effect of intravenous administration of a dialyzed, enzymatic casein hydrolysate (Aminosol) on the serum concentration and on the urinary excretion of amino acids. Peptides and nitrogen. Scand. Lab. clin. Invest. *4:* 167–178 (1952).

Liljedahl, S.-O.: Burn treatment at Karolinska hospital in Stockholm. Opusc. Med. *15:* 179–194 (1970).

Liljedahl, S.-O.: The nutrition of patients with extensive burns; in *Wilkinson* Parenteral nutrition, pp. 208–212 (Churchill, Livingstone, London 1972).

Lindmark, L. and Wretlind, A.: Personal commun. (1975).

Lister, J.: On the effects of the antiseptic system of treatment upon the salubrity of a surgical hospital. Lancet *i:* 4–6, 40–42 (1870).

Long, C.L.; Crosby, F.; Geiger, J.W., and Kinney, J.M.: Parenteral nutrition in the septic patient nitrogen balance limiting plasma amino acids, and calorie to nitrogen ratios. Am. J. clin. Nutr. *29:* 380–391 (1976).

Long, C.L.; Zikria, B.A.; Kinney, J.M., and Geiger, J.-W.: Comparison of fibrin hydrolysates and crystalline amino acid solutions in parenteral nutrition. Am. J. clin. Nutr. *27:* 163–174 (1974).

Long, J.M.; Wilmore, D.W.; Mason, A.D., and Pruitt, B.A.: Fat carbohydrate interaction. Effects on nitrogen sparing in total intravenous feeding. Surg. Forum *25:* 61–63 (1974).

Lower, R. and King, E.: An account of the experiment of transfusion. Phil. Trans. *2:* 557–564 (1662).

MacFadyen, B.V., jr.; Dudrick, S.J., and Ruberg, R.L.: Management of gastrointestinal fistulae with parenteral hyperalimentation. Surgery, St Louis *74:* 100–105 (1973).

Maddock, R.K.; Bloomer, H.A., and St. Jeor, S.W.: Low protein diets in the management of renal failure. Ann. intern. Med. *69:* 1003–1008 (1968).

Maini, B.; Blackburn, G.L.; Bistrian, B.R.; Flatt, J.P.; Page, J.G.; Bothe, A.; Benotti, P., and Rienhoff, H.Y.: Cyclic hyperalimentation: an optimal technique for preservation of visceral protein. J. Surg. Res. *20:* 515–525 (1976).

Major, J.D.: Chirurgia infusoria (Reumannus, Kilonia 1662).

Malcolm, J.D.: The physiology of death from traumatic fever (Churchill, London 1893).

Malvy, P.; Rousseau, C. et Cardon, J.: Utilisation défecteuse de l'alimentation azoté intraveineuse chez des malades chirurgicaux. Accidents consécutifs. Presse méd. *69:* 917–920 (1961).

McGovern, B.: Septic complications of hyperalimentation; in *Cowan and Scheetz* Intravenous hyperalimentation, pp. 165–174 (Lea & Febiger, Philadelphia 1972).

Mendeloff, AL. and Weichselbaum, T.E.: Role of the human liver in the assimilation of intravenously administered fructose. Metabolism *2:* 450–458 (1953).

Mertz, W.: Some aspects of nutritional trace element research. Fed. Proc. Fed. Am. Socs exp. Biol. *29:* 1482–1488 (1970).

Meyer, C.E.; Fancer, J.A.; Schnurr, P.E., and Webster, H.D.: Composition, preparation and testing of an intravenous fat emulsion. Metabolism *6:* 591–596 (1957).

Michel, H.; Raynaud, A.; Crastes de Paulet, P.; Nalet, B.; Orsetti, A. et Bertrand, L.: Tolérance du cirrhotique aux lipides intraveineux; in *Romieu, Solassol, Joyeux et Astruc* Comptes rendus du Congr. Int. de Nutrition Parentérale, pp. 131–137 (Déhan, Montpellier 1976).

Miller, L.L.: The role of the liver and the non-hepatic tissues in the regulation of free amino acid levels in the blood; in *Holden* Amino acid pools, p. 708 (Elsevier, Amsterdam 1962).

Mills, C.F.: Some aspects of trace element nutrition in man. Nutrition *26:* 357–369 (1972).

Moncrief, J.A.: The requirements for parenteral therapy in the burned patient; in *Meng and Law* Parenteral nutrition, pp. 40–44 (Thomas, Springfield 1970).

Moore, F.D.: Metabolic care of the surgical patient (Saunders, Philadelphia 1959).

Moore, F.D. and Ball, M.R.: The metabolic response to surgery (Thomas, Springfield 1959).

Munro, H.N.: Carbohydrate fat as factors in protein utilisation and metabolism. Physiol. Rev. *31:* 449–488 (1951).

Munro, H.N.: General aspects of the regulation of protein metabolism by diet and by hormones; in *Munro and Allison* Mammalian protein metabolism, vol. 1, pp. 381–481 (Academic Press, New York 1964).

Munro, H.N.: Nutritional factors affecting amino acid metabolism; in *Lang, Fekl and Berg* Balanced nutrition and therapy, pp. 1–12 (Thieme, Stuttgart 1971).

Munro, H.N.: Amino acid requirements and metabolism and their relevance to parenteral nutrition; in *Wilkinson* Parenteral nutrition, pp. 34–67 (Churchill, Livingstone, London 1972).

Munro, H.N. and Thompson, W.S.T.: Influence of glucose on amino acid metabolism. Metabolism *2:* 354–361 (1953).

Murlin, J.R. and Riche, J.A.: The fat of the blood in relation to heat production, narcosis and muscular work. Am. J. Physiol. *40:* 146 (1916).

Müller, F.; Strack, E.; Kuhfahl, E. und Dettmer, D.: Der Stoffwechsel von Xylit bei normalen und alloxandiabetischen Kaninchen. Z. ges. exp. Med. *142:* 338–350 (1967).

Najarian, J.S. and Harper, H.A.: Comparative effect of arginine and monosodium glutamate on blood ammonia. Proc. Soc. exp. Biol. Med. *92:* 560–563 (1956).

Obel, A.-L.: Morphological studies in long-term experiments with intravenous fat emulsions in dogs; in *Berg* Advances in parenteral nutrition, pp. 206–216 (Thieme, Stuttgart 1970).

Olivecrona, T.; George, E.P., and Borgström, B.: Chylomicron metabolism. Fed. Proc. Fed. Am. Socs exp. Biol. *20:* 928–933 (1961).

O'Neill, J.A.: Workshop No. 6, delivery; in AMA Symp. Total Parenteral Nutrition, Nashville 1972, pp. 252–257.

Østergaard, J.: Parenteral nutrition of operated patients. Nutr. Diet. *5:* 408–413 (1963).

Owings, J.M.; Bomar, W.E., jr., and Ramage, R.C.: Parenteral hyperalimentation and its practical applications. Ann. Surg. *175:* 712–719 (1972).

Panteliadis, C.; Jürgens, P. und Dolif, D.: Aminosäurenbedarf Früh- und Neugeborener unter den Bedingungen der parenteralen Ernährung. Infusionstherapie klin. Ernähr. *2:* 65–72 (1975).

Pasteur, L. et Joubert, J.-F.: Charbon et septicémie. Compt. r. hebd. Séanc. Acad. Sci., Paris *85:* 101–115 (1877).

Paulsrud, J.R.; Pensler, L.; Whitten, C.F.; Stewart, S., and Holman, R.T.: Essential fatty acid deficiency in infants induced by fat-free intravenous feeding. Am. J. clin. Nutr. *25:* 897–904 (1972).

Paymaster, N.J.: Postoperative magnesium deficiency. Br. J. Anaesth. *47:* 85–87 (1975).

Peaston, M.J.T.: Design of an intravenous diet of amino acids and fat suitable for intensive patient-care. Br. med. J. *ii:* 388–390 (1966).

Peaston, M.J.T.: Maintenance of metabolism during intensive patient care. Post-grad. med. J. *43:* 317–338 (1967).

Peaston, M.J.T.: Parenteral nutrition in serious illness. Hosp. Med. *1968a:* 708–711.

Peaston, M.J.T.: A comparison of *L*- and synthesised *DL*-amino acids for complete parenteral nutrition. Clin. Pharmac. Ther. *9:* 61–66 (1968b).

Peaston, M.J.T.: Protein and amino acid metabolism – response to injury; in *Lee* Parenteral nutrition in acute metabolic illness, pp. 139–166 (Academic Press, New York 1974).

Pendray, M.R.: Peripheral vein feeding in infants: technique, results and problems. Proc. Meet., Vancouver, pp. 158–167 (Pharmacia, Quebec 1974).

Peterkofsky, B. and Udenfriend, S.: Enzymatic hydroxylation of proline in microsomal polypeptide leading to formation of collagen. Proc. natn. Acad. Sci. USA *53:* 335–342 (1965).

Press, M.; Hartop, P.J., and Protley, C.: Correction of essential fatty acid deficiency in man by cutaneous application of sunflower seed oil. Lancet *i:* 597–599 (1974).

Principi, N.; Reali, E., and Rivolta, A.: Sorbitol in total parenteral nutrition in pediatric patients. Helv. paediat. Acta *28:* 621–627 (1973).

Puri, P.; Guiney, E.J., and O'Donnell, B.: Total parenteral feeding in infants using peripheral veins. Archs Dis. Childh. *50:* 133–136 (1975).

Randall, H.T.: Indications for parenteral nutrition in postoperative catabolic states; in *Meng and Law* Parenteral nutrition, pp. 13–39 (Thomas, Springfield 1970).

Recommended Dietary Allowances; 7th ed. (Natn. Academy of Sciences, Washington 1968).

Recommended Intakes of Nutrients for the United Kingdom. Reports on Public Health and Medical Subjects, No. 120 (HMSO, London 1969).

Reid, D.J.: Intravenous fat therapy. II. Changes in oxygen consumption and respiratory quotient. Br. J. Surg. *54:* 204–207 (1967).

Reid, D.J. and Ingram, G.I.C.: Changes in blood coagulation during infusion of Intralipid. Clin. Sci. *33:* 339–407 (1967).

Reinhold, J.G.: Trace elements – a selected survey. Clin. Chem. *21:* 476–500 (1975).

Reissigl, H.: Praxis der Flüssigkeitstherapie (Urban & Schwarzenberg, München 1965).

Reiter, J.M.: Total parenteral alimentation in infants. Northwest Med. *70:* 337–341 (1971).

Rhode, C.M.; Parkins, W.M.; Tourtellotte, D., and Vars, H.M.: Method for continuous intravenous administration of nutritive solutions suitable for prolonged metabolic studies in dogs. Am. J. Physiol. *159:* 409–414 (1949).

Rickham, P.P.: Massive small intestinal resection in newborn infants. Hungarian lecture delivered at the Royal College of Surgeons of England. Ann. R. Coll. Surg. *41:* 480–492 (1967).

Ricour, C.; Gros, J.; Maziere, B., and Comar, D.: Trace elements in infants on total parenteral nutrition. Abstr. 10th Int. Congr. on Parenteral Nutrition, Kyoto 1975a, p. 236.

Ricour, C.; Millot, M., and Balsan, S.: Phosphorus depletion in children on long-term total parenteral nutrition. Acta paediat. scand. *64:* 385–392 (1975b).

Rivers, J.P.W. and Davidson, B.C.: Linoleic acid deprivation in mice. Proc. Nutr. Soc. *33:* 48A–49A (1974).

Robergs, Q.R. and Leung, B.: The influence of amino acids on the neuroregulation of food intake. Fed. Proc. Fed. Am. Socs exp. Biol. *32:* 1709–1719 (1973).

Robison, G.A.; Butcher, R.W., and Sutherland, E.W.: Cyclic-AMP (Academic Press, New York 1971).

Rogers, W.F. and Gardner, F.H.: Tyrosine metabolism in human scurvy. J. Lab. clin. Med. *34:* 1491–1501 (1949).

Rose, W.C.: The significance of the amino acids in nutrition. Harvey Lect. *30:* 49–65 (1934–35).

Rose, W.C.: The amino acid requirements of adult man. Nutr. Abstr. Rev. *27:* 632–647 (1957).

Rose, W.C. and Wixom, R.L.: The amino acid requirements of man. XVI. The role of the nitrogen intake. J. biol. Chem. *217:* 997–1004 (1955).

Rössner, S.: Studies on intravenous fat tolerance test. Methodological, experimental and clinical experiences with Intralipid. Acta med. scand. Suppl. *564:* 1–24 (1974).

Rutten, P.; Blackburn, G.L.; Flatt, J.P.; Hallowell, E., and Cochran, D.: Determination of optimal hyperalimentation infusion rate. J. Surg. Res. *18:* 477–483 (1975).

Sanderson, I. and Deitel, M.: Insulin response in patients receiving concentrated infusions of glucose and casein hydrolysate for complete parenteral nutrition. Ann. Surg. *179:* 387–394 (1974).

Sandstead, H.H.: Present knowledge of the minerals; in Present knowledge in nutrition; 3rd ed., pp. 117–125 (The Nutrition Foundation, New York 1967).

Savege, T.M.: The complications associated with the use of plastic intravenous catheters. Resuscitation *2:* 83–101 (1973).

Schärli, A.: Beitrag zur parenteralen Ernährung bei chirurgischen Kranken. Praxis *53:* 1215–1217 (1964).

Schärli, A.: Praktische Gesichtspunkte bei der vollen parenteralen Ernährung. Int. Z. VitamForsch. *35:* 52–59 (1965).

Schärli, A. und Rumlova, E.: Parenterale Ernährung in der Kinderchirurgie. Ergebnisse von Bilanzuntersuchungen. Infusionstherapie klin. Ernähr. *2:* 51–55 (1975).

Shoefl, G.I.: The ultrastructure of chylomicra and of the particles in an artificial fat emulsion. Proc. R. Soc. Ser. B *169:* 147–152 (1968).

Scholler, K.L.: Transport und Speicherung von Fettemulsionsteilchen. Z. prakt. Anästh. Wiederbel. *3:* 193–194 (1968).

Schroeder, H.A.: The role of chromium in mammalian nutrition. Am. J. clin. Nutr. *21:* 230–244 (1968).

Schuberth, O.: Klinische Erfahrungen mit Fettemulsionen für intravenöse Anwendung. Berl. Med. *14:* 235–236 (1963).

Schuberth, O. and Wretlind, A.: Intravenous infusion of fat emulsions, phosphatides and emulsifying agents. Acta chir. scand. Suppl. *278:* 1–21 (1961).

Schultis, K.: Xylit als Glucoseaustauschstoff bei gestörter Glucoseassimilation im Postaggressions-Syndrom. Eine Übersicht über tierexperimentelle und klinische Studien. Z. ErnährWiss. *10:* suppl. 11, pp. 87–97 (1971).

Schultis, K. and Beisbarth, H.: Posttraumatic energy metabolism; in *Wilkinson* Parenteral nutrition, pp. 255–265 (Churchill, Lvingstone, London 1972).

Schultis, K. and Geser, C.A.: Clinical experiments on the use of carbohydrates in stress; in *Meng and Law* Parenteral nutrition, pp. 139–148 (Thomas, Springfield 1970).

Schumer, W.: Adverse effects of xylitol in parenteral nutrition. Metabolism *20:* 345–347 (1971).

Scribner, B.H.; Cole, J.J.; Christopher, T.G.; Vizzo, J.E.; Atkins, R.C., and Blagg, C.R.: Long-term total parenteral nutrition: the concept of an artificial gut. J. Am. med. Ass. *212:* 457–463 (1970).

Scriver, C.R.; Clow, C.L., and Lamm, P.: Plasma amino acids: screening, quantitation and interpretation. Am. J. clin. Nutr. *24:* 876–890 (1971).

Sedgwick, C.E. and Viglotti, J.: Hyperalimentation. Surg. Clins N. Am. *51:* 681–686 (1971).

Seeling, W.; Ahnefeld, F.W.; Dick, W.; Fodor, L. und Dölp, R.: Die Bedeutung der Spurenelemente im Rahmen einer parenteralen Ernährung am Beispiel der Elemente Kupfer, Zink und Chrom. Z. ErnährWiss. *14:* 302–308 (1975).

Shils, M.E.: Minerals in total parenteral nutrition; in AMA Symp. Total Parenteral Nutrition, 1972, pp. 92–114.

Shils, M.E.: Minerals; in *White and Nagy* Total parenteral nutrition, pp. 257–275 (Urban & Schwarzenberg, Munich 1974).

Shohl, A.T. and Blackfan, K.D.: Intravenous administration of crystalline amino acids to infants. J. Nutr. *20:* 305–316 (1940).

Sinclair, A.J.; Fiennes, R.N.; Hay, A.W.; Watson, G.; Crawford, M.A., and Hart, M.G.: Linolenic acid deprivation in Capruchia monkeys. Proc. Nutr. Soc. *33:* 49A–50A (1974).

Slyke, D.D. v.; McIntosch, J.F.; Muller, E.; Hannon, R.A., and Johnston, C.: Studies of urea excretion. VI. Comparison of the blood urea clearance with certain other measures of renal function. J. clin. Invest. *8:* 357–374 (1930).

Solassol, C. et Joyeux, H.: Nouvelles techniques pour nutrition parentérale chronique. Ann. Anésth. fr. spéc. *II:* 75–85 (1974).

Solassol, C. and Joyeux, H.: Ambulatory parenteral nutrition; in *Fischer* Total parenteral nutrition, pp. 285–301 (Little, Brown, Boston 1976).

Solassol, C.; Joyeux, H.; Serrou, B.; Pujol, H. et Romieu, C.: Nouvelles techniques de nutrition parentérale à long terme pour suppléance intestinale. J. Chir., Paris *105:* 15–24 (1973).

Souchon, E.A.; Copeland, E.M.; Watson, P., and Dudrick, S.J.: Intravenous hyperalimentation as an adjunct to cancer chemotherapy with 5-fluorouracil. J. Surg. Res. *18:* 451–454 (1975).

Steffee, C.H.; Wissler, R.W.; Hamphreys, E.M.; Benditt, E.P.; Woolridge, R.W., and Cannon, P.R.: Studies in amino acid utilization. V. The determination of minimum daily essential amino acid requirements in protein-depleted adult male albino rats. J. Nutr. *40:* 483–497 (1950).

Stegink, L.D. and Besten, L. den: Synthesis of cystine from methionine in normal subjects: effect of route of alimentation. Science *178:* 514–516 (1972).

Steiger, E.; Daly, J.M.; Vars, H.M.; Allen, T.R., and Dudrick, S.J.: Animal research in intravenous hyperalimentation; in *Cowan and Scheetz* Intravenous hyperalimentation, pp. 186–194 (Lea & Febiger, Philadelphia 1972).

Steinbereithner, K.: Problems of artificial alimentation in an intensive therapy unit. Possibilities and limitations; in *Evans and Gray* Modern trends in anaesthesia, chap. 11 (Butterworth, London 1966).

Steinbereithner, K. und Wagner, O.: Das Verhalten des arteriellen Sauerstoffdrucks nach intravenöser Fett- und Laevulosebelastung bei schweren Schädelverletzten. Agressologie *8:* 389–393 (1967).

Stell, P.M.: Esophageal replacement of transposed stomach following pharyngo-laryngoesophagectomy for carcinoma of the cervical oesophagus. Acta oto-lar. *91:* 166–170 (1970).

Stoner, H.B.: Energy metabolism after injury. J. clin. Path. *4:* suppl. 23, pp. 47–55 (1970).

Stoner, H.B. and Heath, D.F.: The effects of trauma on carbohydrate metabolism. Br. J. Anaesth. *45:* 244–251 (1973).

Sturman, J.A.; Gaull, G., and Raiha, N.C.R.: Absence of cystathionase in human fetal liver: Is cystine essential? Science *169:* 74–76 (1970).

Sudjian, A.: Personal commun. (1974).

Sundström, G.; Zauner, C.W., and Arborelius, M., jr.: Decrease in pulmonary diffusing capacity during lipid infusion in healthy men. J. appl. Physiol. *34:* 816–820 (1973).

Swenseid, M.E.; Feeley, R.J.; Harris, C.L., and Tuttle, S.G.: Egg protein as a source of the essential amino acids. Requirement for nitrogen balance in young adults studied at two levels of nitrogen intake. J. Nutr. *68:* 203–211 (1959).

Swift, R.W. and Fischer, K.H.: Energy metabolism; in *Beaton and McHenry* Nutrition. A comprehensive treatise, pp. 181–260 (Academic Press, New York 1964).

Thomas, D.W.; Edwards, J.B., and Edwards, R.G.: Side effects of sugar substitutes during intravenous administration. Nutr. Metabol. *18:* suppl. 1, pp. 227–241 (1975).

Thomas, D.W.; Edwards, J.B.; Gilligan, J.E.; Lawrence, J.R., and Edwards, R.G.: Complications following intravenous administration of solutions containing xylitol. Med. J. Aust. *i:* 1238–1246 (1972).

Thompson, S.W.; Jones, L.D.; Ferrell, J.F.; Hunt, R.D.; Meng, H.C.; Kuyaha, T.; Sasaki, H.; Schaffner, F.; Singleton, W.S., and Cohn, J.: Testing of fat emulsions for toxicity. 3. Toxicity studies with new fat emulsions and emulsion components. Am. J. clin. Nutr. *16:* 43–61 (1965).

Thorén, L.: Parenteral nutrition with carbohydrate and alcohol. Acta chir. scand. suppl. *325:* 75–93 (1963b).

Tolbert, B.M.; Chen, A.W.; Bell, E.M., and Baker, E.M.: Metabolism of 1-ascorbic acid in man. Am. J. clin. Nutr. *20:* 250–252 (1967).

Travis, S.; Sugerman, H.J.; Ruberg, R.L.; Dudrick, S.J.; Delivoria-Papadopoulos, M.; Miller, L.D., and Oski, F.A.: Alterations of red-cell glycolytic intermediates and oxygen transport as a consequence of hypophosphatemia in patients receiving intravenous hyperalimentation. New Engl. J. Med. *285:* 763–768 (1971).

Tweedle, D.E.F. and Johnston, I.D.A.: The effect of environmental temperature and nutritional intake on the metabolic response in abdominal surgery. Br. J. Surg. *58:* 771–774 (1971).

Tweedle, D.E.F.; Spivey, J., and Johnston, I.D.A.: The effect of four different amino acid solutions upon the nitrogen balance of postoperative patients; in *Wilkinson* Parenteral nutrition, pp. 247–254 (Churchill, Livingstone, London 1972).

Underwood, E.J.: Trace elements in human and animal nutrition (Academic Press, New York 1971).

Vlaardingerbroek, V.M.: Experiences with intravenous nourishment. Ned. Tijdschr. Geneesk. *111:* 415–418 (1967).

Voss, U. und Schnell, J.: Parenterale Ernährung mit einer aminosäurehaltigen Fettemulsion im Tierversuch; in *Berg* Advances in parenteral nutrition, pp. 217–221 (Thieme, Stuttgart 1970).

Wadström, L.B. and Wiklund, P.E.: Effect of fat emulsions on nitrogen balance in the postoperative period. Acta chir. scand. Suppl. *325:* 50–54 (1964).

Waterlow, J.C. and Harper, A.E.: Assessment of protein nutrition; in *Ghadimi* Total parenteral nutrition – promises and premises, pp. 231–258 (Wiley & Sons, Chichester 1975).

Wei, P.; Hamilton, J.R., and LeBlanc, A.E.: A clinical and metabolic study of an intravenous feeding technique using peripheral veins as the initial infusion site. Can. med. Ass. J. *106:* 969–974 (1972).

Weismann, K.; Fischer, A., and Hjorth, N.: Zinc depletion syndrome with acrodermatitis during long-term parenteral nutrition. Two cases treated with oral and intravenous zinc (in Danish). Ugeskr. Laeg. *138:* 1403–1406 (1976).

Weiss, Y. and Nissan, S.: A method for reducing the incidence of infusion phlebitis. Surgery Gynec. Obstet. *141:* 73–74 (1975).

Welch, C.E. and Edmunds, L.H.: Gastrointestinal fistulas. Surg. Clins N. Am. *42:* 1311–1320 (1962).

Wilmore, D.W.: Evaluation of the patient; in AMA Symp. Total Parenteral Nutrition, Nashville 1972, pp. 239–247 (1972).

Wilmore, D.W. and Dudrick, S.J.: Treatment of acute renal failure with intravenous essential *L*-amino acids. Archs Surg., Chicago *99:* 669–673 (1969a).

Wilmore, D.W. and Dudrick, S.J.: Safe long-term venous catheterization. Archs Surg., Chicago *98:* 256–258 (1969b).

Wilmore, D.W. and Dudrick, S.J.: Effects of nutrition in intestinal adaptation following massive small bowel resection. Surg. Forum *20:* 398–400 (1969c).

Wilmore, D.W.; Groff, D.B.; Bishop, H.C., and Dudrick, S.J.: Total parenteral nutrition in infants with catastrophic gastrointestinal anomalies. J. pediat. Surg. *4:* 181–189 (1969).

Wilmore, D.W.; Moylan, J.A.; Helmkamp, G.M., and Pruitt, B.A., jr.: Clinical evaluation of a 10% intravenous fat emulsion for parenteral nutrition in thermally injured patients. Ann. Surg. *178:* 503–513 (1973a).

Wilmore, D.W.; Moylan, J.A.; Landsay, C.A.; Faloona, G.R.; Unger, G.R., and Pruitt, B.A.: Hyperglucagonemia following thermal injury. Insulin and glucagon in the posttraumatic catabolic state. Surg. Forum *24:* 99–101 (1973b).

Witzel, L.; Berg, G.; Grabner, W. und Bergner, D.: Parenterale Ernährung mit einer kombinierten Fettkohlenhydrat-Aminosäure-Lösung. Med. Ernähr. *11:* 177–179 (1970).

Wolf, H.; Melichar, V.; Berg, W., and Kerstar, J.: Intravenous alimentation with a mixture of fat, carbohydrates and amino acids in small immature newborn infants – a preliminary report. Infusionstherapie klin. Ernähr. *1:* 479–481 (1973).

Woods, H.F.: Energy provision during intravenous feeding. Clin. Trials J. *12:* suppl. 1, pp. 62–68 (1975).

Woodyatt, R.T.; Sansum, W.D., and Wilder, R.M.: Prolonged and accurately timed intravenous injections of sugar. J. Am. med. Ass. *65:* 2067–2070 (1915).

Wren, C.: in *Annan* An exhibitation of books on the growth of our knowledge of blood transfusion. Bull. N.Y. Acad. Med. *15:* 622 (1939).

Wretlind, A.: Free amino acids in dialyzed casein digest. Acta physiol. scand. *13:* 45–54 (1947).

Wretlind, A.: Complete intravenous nutrition. Theoretical and experimental background. Nutr. Metabol. Suppl. *14:* 1–57 (1972).

Wretlind, A.: Fat; in *Ghadimi* Total parenteral nutrition – premises and promises, chap. 3, pp. 23–46 (Wiley & Sons, Chichester 1975).

Wretlind, A. and Roos, K.-A.: Studies on intravenous nutrition in rats. Abstr. 10th Int. Congr. of Nutrition, Kyoto 1975, p. 237.

Yamakawa, S.: Nippon Naika Gakkai Zasshi *17:* 122 (1920).

Yokoyama, K.; Okamoto, T.; Tsuda, Y., and Suyama, T.: Metabolism of intravenously injected fat emulsion. Abstr. 10th Int. Congr. of Nutrition, Kyoto 1975, p. 226.

Yoshimura, N.N.; Ehrlich, H.; Westman, T.L., and Deindoerfer, F.H.: Maltose in total parenteral nutrition of rats. J. Nutr. *103:* 1256–1261 (1973).

Young, C.M.; Scanlan, S.S.; Im, H.S., and Lutwak, L.: Effect on body composition and other parameters in obese men of carbohydrate level of reduction diet. Am. J. clin. Nutr. *24:* 290–296 (1971).

Young, J.M. and Weser, E.: The metabolism of circulatory maltose in man. J. clin. Invest. *50:* 986–991 (1971).

Young, V.R.; Hussein, M.A.; Murray, E., and Scrimshaw, N.S.: Plasma tryptophan curve and its relation to tryptophan requirements in young adult men. J. Nutr. *101:* 45–59 (1971).

Zumtobel, V. und Zehle, A.: Postoperative parenterale Ernährung mit Fettemulsionen bei Patienten mit Leberschäden. Arch. klin. Chir. Suppl. Chir. Forum, pp. 179–182 (1972).

Dr. *Alan Shenkin,* Department of Biochemistry, Royal Infirmary, *Glasgow G4 0SF* (Scotland)
Prof. *Arvid Wretlind,* VIHN, Vitrum Institute for Human Nutrition, *102 24 Stockholm 12* (Sweden)

Wld Rev. Nutr. Diet., vol. 28, pp. 112–142 (Karger, Basel 1978)

Biochemistry and Physiology of Magnesium

Jerry K. Aikawa

University of Colorado School of Medicine, Denver, Colo.

Contents

I. Introduction

Magnesium is one of the most plentiful elements on earth; in the vertebrate, it is the fourth most abundant cation. Magnesium is associated with so many different biological processes that this involvement suggests that it has some single fundamental role (4). It is the purpose of this review to summarize the current knowledge of the biochemistry and physiology of magnesium.

II. Biochemistry of Magnesium

A. The Role of Magnesium in Photosynthesis

1. Chloroplast

One of the greatest triumphs of early evolution was the invention of a means to harness the energy of the sun which is transmitted as light to drive energy-requiring synthetic processes. This process in higher plants occurs in an especially organized subcellular organelle, the chloroplast. The chloroplast is an organized set of membranes, crowded with water-insoluble lipid and containing the central pigment chlorophyll. Chlorophyll is the *magnesium* chelate of porphyrin.

2. Chlorophyll

It is chlorophyll which produces the oxygen and the foods for all other forms of life on earth. The excess production of oxygen soon made the oxidation of organic compounds thermodynamically favorable. Under these new conditions, the desired thermodynamically uphill reactions would be the photoreduction of the oxidized organic compounds, including carbon dioxide. It is just these photoreactions which are favored by chelating a dipositive closed shell metal ion into the porphyrin ring. *Mauzerall* (94) hypothesizes that the purpose of the ionically bound magnesium in the vacant hole in the porphyrin ring is to stabilize the structure so that it would undergo perfectly reversible one-electron oxidations. The redox potential of chlorophyll correlates very well with the electronegativity of the central magnesium ion. With photoactivation, the chelated magnesium makes the excited state a powerful reductant and stabilizes the resulting cation. Why magnesium rather than some other metal in the photosynthetic pigments? *Fuhrhop and Mauzerall* (47) suggest that if the aim of the biological system is a minimum redox potential combined with maximum stability in a protonic solvent, then magnesium is a good minimax solution to the requirements.

3. Regulation of Photosynthesis by Magnesium

One can imagine that the chlorophyll molecule harvests light in the manner of a lightmeter. The light quantum activates the chlorophyll molecule, i.e. an electron moves from the π orbitals to the exterior of an atomic shell and then is ejected, leaving behind a chlorophyll-free radical. This terminates the true photochemical event, in which a light quantum is transmuted into a high-energy electron. *Lin and Nobel* (79) observed an increase in the concentration of chloroplast Mg^{++} *in vivo* caused by illuminating the plant, the first direct evidence indicating that changes in magnesium level actually occur in the plant cell. This extra magnesium in the chloroplasts enhanced the photophosphorylation rate. Thus, the increase in magnesium in chloroplasts may be a regulatory mechanism whereby light controls photosynthetic activity.

The energy of electrons is used to produce the adenosine triphosphate (ATP) which, together with reduced nicotinamide adenine dinucleotide (NADPH), drives the formation of carbohydrates from CO_2. Therefore, the chloroplast is a transducer which converts the electromagnetic energy from the sun into the chemical energy of ATP. This transduction does not occur in the absence of the chelated magnesium.

The chloroplast is located physically in the granum. The granum possesses a lamellar structure which is compatible with the existence of an interface between hydrophilic and hydrophobic phases, in which the chlorophyll molecule could be completely accommodated in a closely packed or monomolecular layer. From this characteristic lamellarity evolves the concept of a unit membrane held together with the assistance of magnesium. Magnesium could stiffen lipoprotein membranes by bridging neighboring carboxylate groups (66). Chlorophyll functions in photosynthesis by virtue of its ability to produce and to maintain a charge separation in the highly ordered lamellar structure of the chloroplast.

4. A View of Photosynthesis

The entire photosynthetic process can be viewed as the capture of the energy of photons in the form of high-energy electrons, followed by a stepwise passage of electrons down an energy gradient in a structured membrane held together by the coordinating properties of the magnesium atom. This model is essentially the unifying concept first proposed by *Szent-Györgi* (124); it is now modified to explain the role of magnesium.

The subsequent synthetic process may be summarized as follows: $6CO_2 + 6H_2O + 18\ ATP + 12\ NADPH \rightarrow C_6H_{12}O_6 + 18\ ADP + 18\ Pi + 12\ NADP + 6O_2$. ATP and chemical reducing power operate to produce carbohydrate via common intermediates which require twelve separate enzymes. All enzyme reactions that are known to be catalyzed by ATP show an absolute requirement for magnesium.

B. *The Role of Magnesium in Oxidative Phosphorylation*

In the absence of sunlight, plants rely on stored chemical energy to maintain life. This stored energy is released by oxidative phosphorylation, a process which occurs in the mitochondrial membrane of both plants and animal cells. The primary function of all mitochondria is to couple phosphorylation to oxidation. ATP, the main fuel of life, is produced in oxidative phosphorylation. All enzyme reactions that are known to be catalyzed by ATP show an absolute requirement for magnesium. These reactions encompass a very wide spectrum of synthetic processes.

So fundamental and widespread are the reactions involving ATP, that it must influence practically all processes of life. Many enzymes are activated by the magnesium cation; this group includes all those utilizing ATP or catalyzing the transfer of phosphate. ATP is known to form a magnesium complex, with Mg^{++} binding usually to the phosphate moiety. Magnesium has a single divalent state and does not form highly stable chelates with organic complexes, as do the transitional metals. It is perhaps this quality which allows it to act as a bridge in a large number of chemical reactions not requiring redox reactions, but resulting in transfer of organic groups from one molecule to another. When organic phosphate takes part in a reaction, magnesium is usually its inorganic cofactor. All partners in reactions known to be dependent on ATP are capable of chelating with magnesium. The effect of magnesium chelation in such reactions is to lower the free energy of activation of the rate-determining step.

The ATP molecule is usually depicted as existing in a linear configuration, with the purine and the phosphate ends separated by the pentose. *Szent-Györgi* (125) has suggested that the spacial configuration of the ATP molecule is such that it could function as a transformer as well as a storage battery. The phosphate chain can touch the purine ring; magnesium can form a very stable quadridentate chelate connecting the two ends of the ATP molecule, and energy in the form of electrons can now pass from the phosphate to the purine. The magnesium may not only actually connect the two ends of the molecule, but it may also make one single, unique electronic system of the phosphate chain and the purine with common nonlocalized electrons which could transport energy.

C. The Mitochondrion

The role of sunlight, chlorophyll, and magnesium in the primary synthetic process on earth have already been discussed; in photosynthesis, carbon dioxide and water are synthesized into carbohydrate, and oxygen is released. In the absence of sunlight, plants rely on stored chemical energy to maintain life. This stored energy is released by oxidative phosphorylation, a process which occurs in the mitochondrion of both plant and animal cells. The biosynthesis of ATP coupled to the oxidation of substrate is known as oxidative phosphorylation and takes place in the mitochondrial membrane. The primary function of all mitochondria is to couple phosphorylation to oxidation. The transduction is the conversion of chemical energy from the bond energies of certain metabolites to the bond energies of ATP. Whereas phosphorylation in the chloroplast is light-dependent, phosphorylation in the mitochondrion is dependent not on light but on oxygen. Whereas photosynthesis combines carbon dioxide and water and evolves oxygen, oxidative phosphorylation does just the reverse. It requires oxygen and evolves carbon dioxide and water, thus completing the carbon cycle on earth and returning the electrons to ground state.

ATP, the main fuel of life, is produced in both photosynthesis and oxidative phosphorylation. In both cases, ATP is produced by an electric current, i.e. the energy released by 'dropping' electrons.

The mitochondrion represents a general blueprint that is characteristic of all membrane systems; in fact, it is characteristic of all the energy-transforming systems of the cell (56). The basic design of the mitochondrion is copied in all other systems in the cell that have to do with the transformation or use of energy. Under the electron microscope, the mitochondrion, just like the chloroplast, is seen to consist of a lamellar membrane. The inner membrane forms invaginations (cristae). The intermembrane and intercristal spaces are thought to be continuous and to form a central compartment. The intermembrane compartment of rat liver mitochondria contains high molecular weight compounds, most likely proteins, which form complexes with magnesium ions (22). The matrix, which is surrounded by the folded inner membrane, comprises the second compartment, and the entire mitochondrion is thought to be a two-compartment system. The cristae contain a strictly regulated respiratory chain along which electrons are transferred by the difference in redox potentials. Along this respiratory chain, the oxidation-reduction energy is converted into phosphate bond energy in the form of ATP. The optimum concentration of magnesium for the process appears to be 10^{-4} to 10^{-5} M. The respiratory enzymes — cytochromes and flavoproteins — which sequentially release the energy of the electrons, may be embedded in the mitochondrial membranes that are structurally organized into respiratory units. The oxidizing enzymes in the inner mitochondrial membrane are assembled asymmetrically in a way that gives rise to a vectorial movement of protons. *Racker* (106) feels that this proton current is the driving force in the production of biologically useful energy.

The traditional concept of a mitochondrion is that of small, discrete, intracellular organelles, relatively free in the cytoplasm. Recent serial section studies indicate that in the yeast (65) and in the rat liver (23) there is but one mitochondrion per cell, consisting of a single branching tubular structure.

Magnesium is present inside the mitochondrial membrane at a concentration of about 1 nmol/mg protein and plays an essential regulatory role in the maintenance of membrane integrity; the presence of magnesium on a number of membrane sites appears to be necessary to maintain the impermeability of the mitochondrial inner membrane (21). Specific pathways for electrophoretic penetration of monovalent cations are present in the inner membrane of the mitochondrion; Mg^{2+} bound by a limited number of high-affinity sites in or near these pathways can control monovalent cation permeability (134). The mitochondrion can be made to swell and contract experimentally. Although swelling can be caused by a large variety of different chemical agents, it is significant that only ATP together with magnesium can cause contraction. ATP is always split during contraction of swollen mitochondria.

A rapid swelling of heart and liver mitochondria can be produced in rats fed a magnesium-deficient diet for 10 days. whereas no significant decrease in the magnesium content of the mitochondrion results. ATP reverses the swelling of mitochondria from heart and liver of magnesium-deficient rats.

Life could have been no more than an experiment of nature until protoorganisms developed dependable machinery to perform two basic functions: (a) generate energy in a form usable for the organisms' various requirements (ATP), and (b) reproduce themselves. It is of considerable theoretical interest that all forms of life on earth have basically the same system for these purposes. They are summed up in the familiar initials ATP and DNA (deoxyribonucleic acid). The relationship between ATP and magnesium has already been discussed. We shall discuss next the involvement of magnesium in the biochemistry of DNA.

D. Magnesium and DNA

During the past 20 years, scientists have obtained substantial understanding of how information is stored and replicated in DNA molecules, how it is passed on to RNA molecules and finally to proteins, and how the three-dimensional structure of proteins depends upon the linear arrangement of the constituent amino acids. This information storage in molecular structure and its subsequent readout is dependent upon the presence of magnesium in optimal concentration (74). The rather complicated three-dimensional structures assumed by some polymers are a consequence of their primary sequence. The interactions of these to form even more complicated multicomponent complexes are also determined by their chemistry.

One of the most important chemical constituents of the cell is DNA, which is almost exclusively confined to the cell nucleus. DNA is the carrier of genetic information.

Much of the magnesium in the cell nucleus is combined with those phosphoric groups of DNA which are not occupied by histone. The chemical factors that control the variable activity at the sites along a chromosome are largely unknown. There is a suggestion that the sites along the DNA chain at which the phosphoric acid groups are combined with histone are inactive and, conversely, that those at which they are combined with magnesium are active. The physical integrity of the DNA helix appears to be dependent upon magnesium. There is evidence to suggest that Mg^{++} is necessary as an intermediate complexing agent during cell duplication and during the formation of ribonucleic acid (RNA) on a double-stranded DNA template.

Both magnesium and ATP are involved in the synthesis of nucleic acids. Since sections of the chromosomes in the nucleus are held together by calcium and magnesium, it seems likely that changes in the concentration of magnesium in the medium might determine the degree of chromosomal aberration. There is evidence that variations in the concentration of magnesium *in vivo* exerts control on DNA synthesis.

E. Magnesium and the Ribosome

Ribosomes are of universal occurrence in microorganisms, higher plants, and animals. The principal and probably the only function of the ribosome is the biosynthesis of protein. The rate of protein synthesis is proportional to the number of ribosomes present. Ribosomes require magnesium ions in order to maintain their physical stability (66); they dissociate into smaller particles when the magnesium concentration becomes low (140). An optimum intracellular concentration of magnesium is required for the integrity of the macromolecules necessary for RNA synthesis (34). The physical size of the RNA aggregates is controlled by the concentration of magnesium, and polypeptide formation cannot proceed unless magnesium concentration is optimal (137). Mg^{++} probably acts to stabilize a favorable protein conformation (30).

F. Discussion

There is very little doubt that magnesium is essential for life on earth. The exact function of magnesium in the chlorophyll molecule, as well as in maintaining the structural integrity of the granum, is conjectural; however, it seems possible that magnesium, because of its inherent atomic composition is, in this particular situation, able to capture and transmit energy more efficiently than any other element. Moreover, the magnesium atom is able to hold reacting groups together and to thus maintain the physical configurations that are optimal either for the transfer of energy in the form of excited electrons or for the transmutation of energy into ATP. The one fundamental property of magnesium upon which all of these photosynthetic processes depends is *chelation*. It seems that the capture, conversion, storage, and utilization of solar energy are all dependent upon a chelating function which is unique to, and specific for, the magnesium atom.

Two of the basic functions of solar energy in living cells are *genetic transcription* and *protein synthesis*. Recent studies in molecular biology have established that interrelations exist among the three major biologic macromolecules: DNA, RNA, and proteins. Genetic information stored in DNA is transcribed into messenger RNA, which in turn translates that information into amino acid sequences in the newly synthesized protein. At literally every turn in these processes, magnesium plays a vital role. The physical integrity of the DNA helix appears to be dependent upon magnesium. The physical size of the RNA aggregates is controlled by the concentration of magnesium, and polypeptide formation cannot proceed unless magnesium concentration is optimal. Magnesium appears to play a central role in the coordinate control of growth and metabolism in animal cells (110).

Rasmussen (108) has recently discussed the role of calcium in the 'closed-loop' feedback system necessary for the mediation of hormonal action; he predicts that it is likely that Mg^{++} will prove to be another divalent cation with a messenger function as complex as that of calcium. This messenger function for magnesium may involve chemical binding to establish physical proximity of reacting groups.

III. Physiology of Magnesium

A. Normal Distribution and Turnover of Magnesium in Man

1. Body Content

The limited data available from analysis of human carcasses indicate that the magnesium content of the human body ranges between 22.7 and 35.0 mEq/kg wet weight of tissue (135). Extrapolations from tissue analyses performed on victims of accidental death indicate that the body content of magnesium for a man weighing 70 kg would be on the order of 2,000 mEq (24 g) (112).

89% of all the magnesium in the body resides in bone and muscle. Bone contains about 60% of the total body content of magnesium at a concentration of about 90 mEq/kg wet weight. Most of the remaining magnesium is distributed equally between muscle and nonmuscular soft tissues. Of the nonosseus tissues, liver and striated muscle contain the highest concentration, 14—16 mEq/kg. Approximately 1% of the total body content of magnesium is extracellular. The levels of magnesium in serum of healthy people are remarkably constant, remaining on the average of 1.7 mEq/l, and varying less than 15% from this mean value (130). The distribution of normal values for serum magnesium is identical in men and women and remains constant with advancing age (71). Approximately one third of the extracellular magnesium is bound nonspecifically to plasma proteins. The remaining 65% which is diffusible or ionized, appears to be the biologically active component. The ratio of bound to unbound magnesium, as well as the total serum levels, is remarkably constant. The magnesium content of erythrocytes varies from 4.4 to 6.0 mEq/l (19).

2. Intake

The average American ingests daily between 20 and 40 mEq of magnesium; magnesium intakes of from 0.30 to 0.35 mEq/kg/day are thought to be adequate to maintain magnesium balance in normal adult (69). A daily intake of 17 mEq (0.25 mEq/kg) may meet nutritive requirements provided that the individual remains in positive magnesium balance. *Schroeder et al.* (112) called attention to the theoretic relationship of dietary magnesium deficiency to serious chronic diseases, including atherosclerosis. The estimated daily requirements for a child is 12.5 mEq (150 mg) (36). The greater importance of magnesium in childhood is suggested by the relative ease with which deficiency states are produced experimentally in young animals as compared with adult animals (36).

Some common foods can be ranked in order of decreasing mean concentrations of magnesium, as follows: nuts, 162 mEq/kg; cereals, 66; sea foods, 29; meats, 22; legumes, 20; vegetables, 14; dairy products, 13; fruits, 6; refined sugars, 5, and fats, 0.6. This order differs when the concentrations are ranked on the basis of the caloric values of the foods, as follows: vegetables, legumes, sea foods, nuts, cereals, dairy products, fruit, meat, refined sugars, and fats. Noteworthy is the very small contribution of fats and refined sugars to the total intake of magnesium. These two, the major sources of caloric energy, are virtually devoid of magnesium (112).

3. Absorption

When a tracer dose of ^{28}Mg was administered orally to 26 subjects, fecal excretion within 120 h accounted for 60—80% of the administered dose (1). The concentration of radioactivity in the plasma was maximal at 4 h, but the actual increase in serum magnesium concentration was negligible. When ^{28}Mg was injected intravenously into a normal human subject, only 1.8% of the radioactivity was recovered in the stool within 72 h (5). The fecal magnesium appears to be primarily magnesium from material that is not absorbed by the body rather than magnesium secreted by the intestine. Ingested magnesium appears to be absorbed mainly by the small intestine (112). The factors controlling the gastrointestinal absorption of magnesium are poorly understood.

4. Secretion

There undoubtedly is considerable secretion of magnesium into the intestinal tract from bile and from pancreatic and intestinal juices. This secretion is followed by almost complete reabsorption. Parotid saliva contains about 0.3 mEq/l (75) and pancreatic juice about 0.1 mEq/l of magnesium. The concentration of magnesium in other secretions varies considerably. The observation that hypomagnesemia can occur in patients suffering from large losses of intestinal fluids suggests that intestinal juices contain enough magnesium to deplete the serum when magnesium is not reabsorbed by the colon.

Studies are just beginning on the role played in the transport of divalent cations by biochemical changes in the cells of the intestinal mucosa (87). Further investigations may show that the cells of the intestinal mucosa, like those in the kidney and elsewhere in the body, may depend in part upon metabolic activity for the uptake and release of calcium and magnesium.

5. Excretion

Most of that portion of the magnesium which is *absorbed* into the body is excreted by the kidney; fecal magnesium represents largely the unabsorbed fraction. In subjects on a normal diet, one third or less of the *ingested* magnesium (5–17 mEq) is excreted by the kidney. After the intravenous injection of a tracer dose of ^{28}Mg in 12–16 mEq of stable magnesium, the daily urinary excretion of magnesium in 8 normal subjects ranged between 6 and 36 mEq (5). Urinary excretion increased as the parenteral dose was increased. The maximal renal capacity for excretion is not known, but it is probably quite high, perhaps greater than 164 mEq/day (130).

The diffusible magnesium in plasma is filtered by the glomeruli and is reabsorbed by the renal tubules, probably by an active process, although the control mechanisms are not known. There is some evidence that magnesium may be secreted by the renal tubule (46). Both the mercurial and the thiazide diuretics increase excretion of magnesium, calcium, potassium, and sodium.

Magnesium excretion also occurs in sweat (35). When men are exposed to high temperature for several days, from 10 to 15% of the total output of magnesium is recovered in sweat. Acclimatization does not occur, as in the case for sodium and potassium. Under extreme conditions, sweat can account for 25% of the magnesium lost daily; this factor would be important when the intake of magnesium is low.

6. Magnesium Conservation on a Low-Magnesium Diet

It is primarily the ionic fraction of the magnesium in plasma which appears in the glomerular filtrate. Any protein-bound magnesium which is filtered is probably returned to the circulation via lymph. The excretion of magnesium may be greater than normal in renal diseases associated with heavy proteinuria.

Magnesium clearance, corrected for protein binding, increases as a linear function of serum magnesium concentration and approaches the inulin clearance at high plasma levels of magnesium. There normally appears to be almost maximal tubular reabsorption of magnesium (32).

In spite of the probability of diets being low in magnesium under certain circumstances, magnesium deficiency does not occur in human beings with healthy kidneys. The explanation for this clinical observation appears to be that renal mechanisms are efficient enough to conserve all but about 1 mEq of magnesium/day. Fecal losses are minimal (17).

7. Abnormal Magnesium Levels in the Blood

Values lower than 1.1 mEq/l have been obtained in patients with congestive heart failure, cirrhosis, or renal failure after hemodialysis. All values higher than 2.0 mEq/l were found in patients with renal failure before therapy (1).

B. The Plasma Clearance and Tissue Uptake of Magnesium

1. Early Studies

Mendel and Benedict (98) reviewed much of the early literature on the absorption and excretion of magnesium. These investigators showed quite clearly that rapid renal excretion of magnesium followed the subcutaneous injection of various magnesium salts, whereas intestinal excretion was minimal. *Hirchfelder and Haury* (64), however, reported that in seven normal adults, 40—44% of an injected dose of magnesium appeared in the urine within 24 h. *Tibbetts and Aub* (128), by means of classic balance techniques, studied the excretion of magnesium in normal subjects; they found that individuals on an oral intake of 49—74 mEq/day excreted 41—66 mEq, of which slightly over one half was in the stools. *Smith et al.* (118) studied the excretion of magnesium in dogs after the intravenous administration of $MgSO_4$ and concluded that the magnesium distributed itself throughout the extracellular fluid during the first 3—4 h; during subsequent hours, some of the ion appeared to be segregated from the extracellular fluid and not excreted.

2. Tracer Studies in Human Beings

The introduction of the radioactive isotope of magnesium, ^{28}Mg, for clinical studies in 1957 made possible determination of the 'exchangeable' pool in human subjects. When nine normal subjects were given intravenous infusions of 12–30 mEq of magnesium tagged with ^{28}Mg, the material was very rapidly cleared from the extracellular fluid (5). The concentration of radioactivity in plasma and urine was too low to follow beyond 36 h. Within a few hours, the volume of fluid available for the dilution of this ion, as calculated from the plasma concentration of ^{28}Mg, exceeded the volume of total body water.

The clearance curves in general showed a rapid phase during the first 4 h, a subsequent more gradual decline up to about 14 h, and a slow exponential slope thereafter. Biopsies of tissues contained concentrations of ^{28}Mg in liver, appendix, fat, skin, and subcutaneous connective tissue which could not be attributed solely to the extracellular components of these tissues. All of these observations suggested that ^{28}Mg rapidly entered cells of the soft tissues and that 70% or more of the infused magnesium was retained in the body for at least 24 h.

Of interest is the observation that the 24-hour urinary excretion of stable magnesium following the infusion of ^{28}Mg approximated the amount of non-radioactive magnesium infused, whereas only 20% of the ^{28}Mg infused was recovered. Previous investigators without the benefit of the radioisotopic data have assumed that most of the infused magnesium was rapidly excreted by the kidney. The additional isotopic data indicate that the infusion of fairly large amounts of magnesium results in a compensatory renal excretion of the body store of magnesium and that the material excreted is probably not the ions that were administered.

Serial external surveys of radioactivity over the entire body revealed the maximal distribution of radioactivity at the end of infusion over the right upper quadrant of the abdomen. This finding suggests initial concentration of magnesium in the liver. At 18 h, the specific activity in bile was equal to that of serum. This equilibration of the infused ^{28}Mg had occurred earlier in bile than in any other tissue or fluid available for study (5).

After about 18 h, the specific activities in plasma and urine showed only a slight gradual increase, suggesting that the infused material had equilibrated with the stable magnesium in a rather labile pool and that further exchange was occurring very slowly in a less labile pool. The size of this labile pool in normal subjects ranged between 135 and 397 mEq (2.6–5.3 mEq/kg of body weight). Since the body content of magnesium is estimated to be 30 mEq/kg, it appears that less than 16% of the total body content of magnesium is measured in the ^{28}Mg exchange technique.

The results of the external survey and the tissue analyses suggest that the labile pool of magnesium is contained primarily in connective tissue, skin, and the soft tissues of the abdominal cavity (such as the liver and intestine) and that the magnesium in bone, muscle and red cells exchanges very slowly.

In another study, *Silver et al.* (117) followed the turnover of magnesium for periods up to 90 h after ^{28}Mg was injected intravenously into human subjects. Even at 90 h, only one third of the body's magnesium had reached equilibrium with the isotope. The results confirmed the impression that the gastrointestinal absorption of magnesium is very limited. Graphic analysis of urinary ^{28}Mg curves in terms of exponential components yielded a slow component with a half-time of 13—35 h, which accounted for 10—15% of the injected dose, and two more rapid components with half-time of 1 and 3 h each, accounted for 15—25% of the injected dose. The large fraction remaining — about 25—50% of the body's total — had a turnover rate of less than 2% per day. Because approximately 25—50% of the total body content exchanges at a turnover rate of less than 2% per day, this isotopic dilution method, used so successfully with sodium and potassium, cannot be employed to quantitate the total body content of magnesium in man. In rabbits, however, the exchangeable magnesium value at 24 h agrees well with the total carcass content of magnesium (7). During starvation, the renal excretion of magnesium amounts to 61.7 mEq/kg of weight loss (2).

3. Magnesium Equilibration in Bone

The reactivity of the skeleton, as measured by isotopic exchange, declines with age (24). The exchange of ^{28}Mg, expressed as bone/serum-specific activity, is much more rapid in younger animals than in older ones. ^{28}Mg accumulates in the bones of young rats about twice as fast as in the bones of adult rats (76).

The exchange of ^{28}Mg in cortical bone occurs much more rapidly in young rats than in old ones. The stable magnesium content of bone increases with age and varies inversely with the water content of bone. ^{28}Mg studies in lambs indicate that the magnesium reserve in bone is mobilized during dietary magnesium deficiency (95).

4. ^{28}Mg Compartmental Analysis in Man

Avioli and Berman (15) used a combination of metabolic balance and ^{28}Mg turnover techniques in order to develop a mathematical model for magnesium metabolism in man. The data thus derived were subjected to compartmental analysis using digital computer techniques.

After the intravenous administration of ^{28}Mg, the decline in the specific activity of plasma or urine can be expressed as the sum of several exponential terms by the method of graphic analysis. On the basis of such analyses, *Silver et al.* (117) defined in man three exchangeable magnesium compartments with half-times of 38, 3, and 1 h. *MacIntyre et al.* (86) described three exchangeable magnesium compartments containing 7.3, 24.4, and 98.7 mEq of magnesium. *Zumoff et al.* (141) obtained similar data.

Multicompartmental analysis indicates that in man there are at least three exchangeable magnesium pools with varied rates of turnover: compartments 1 and 2, exemplifying pools with a relatively fast turnover, together approximating extracellular fluid in distribution; compartment 3, an intracellular pool containing over 80% of the exchanging magnesium with a turnover rate of one half that of the most rapid pool; and compartment 4, which probably accounts for most of the whole-body magnesium. Only 15% of whole-body magnesium, averaging 3.54 mEq/kg body weight, is accounted for by relatively rapid exchange processes (15).

C. Gastrointestinal Absorption

1. Daily Absorption in Man

In normal individuals on regular diets, the average daily absorption of magnesium from the gastrointestinal tract is 0.14 mEq/kg, an amount approximately 40% of the size of the extracellular pool. The rate of entry of magnesium into the intracellular pool would be approximately 0.0058 mEq/kg/h if one assumes that absorption occurs continuously throughout the day. This rate of entry is approximately 1% of the rate of removal of magnesium from the extracellular pool by all routes (132).

2. Factors Affecting Absorption

No single factor appears to play a dominant role in the absorption of magnesium as does vitamin D in the absorption of calcium. Several studies using ^{28}Mg suggest that the absorption of magnesium in man is influenced by the load presented to the intestinal mucosa (6, 55). On an ordinary diet containing 20 mEq of magnesium, 44% of the ingested radioactivity was absorbed per day. On a low-magnesium diet (1.9 mEq/day), 76% was absorbed. On a high-magnesium diet (47 mEq/day), absorption was decreased to 24%.

Absorption begins within an hour of ingestion and continues at a steady rate for 2–8 h; it is minimal after 12 h. In man, absorption throughout the small intestine is fairly uniform, but little or no magnesium is absorbed from the large bowel (55).

3. Site of Absorption

Evidence from a variety of animals suggests that the small intestine is the main site of magnesium absorption, but that the pattern of absorption varies with the species studied (42, 55). Absorption from the large intestine is negligible in the rabbit (7). In male albino rats, more than 79% of the total absorption of ^{28}Mg takes place in the colon, and excretion of endogenous magnesium occurs predominatly in the proximal gut (33). Both magnesium and calcium are bound to phosphate and to nonphosphate-binding material of an unknown nature in the ileal contents of ruminating calves (119), and hence are rendered nonultrafiltrable.

There appears to be an interrelationship between the absorption of magnesium and calcium in the proximal part of the small intestine in the rat (8). The suggestion has been made that there is a common mechanism for transporting calcium and magnesium across the intestinal wall (62, 83).

4. The Role of Ionic Magnesium

At the present time, there is no unequivocal evidence that magnesium is actively transported across the gut wall (3). It seems reasonable to assume that the net amount of dietary magnesium absorbed is directly related to the intake and to the time available for absorption of the magnesium from the small intestine. Therefore, apart from a small effect from the difference in potential across the wall of the small intestine, the concentration of *ionic* magnesium in the digest at the absorption site must be the main factor controlling the amount absorbed in a given time (119).

D. Renal Excretion

1. Control of Body Content

The kidney is the major excretory pathway for magnesium once it is absorbed into the body (98). In subjects on a normal diet, this renal excretion amounts to one third or less of the 5–17 mEq of magnesium which is ingested every day. The mean daily excretion of magnesium in the urine of 12 normal men on an unrestricted diet was 13.3 ± 3.5 mEq (130). Following the intravenous injection of a tracer dose of ^{28}Mg in 12–16 mEq of stable magnesium, the daily urinary excretion of magnesium in eight normal subjects ranged between 6 and 36 mEq (6). Urinary excretion of magnesium increased as the parenteral dose was increased.

Metabolic balance studies in 27 subjects on a self-selected diet of normal composition showed a close positive correlation between the level of dietary intake and the magnesium excretion in both the urine and the feces (61). These results suggest that the absorption of magnesium from the intestinal tract is a poorly controlled process which is determined largely by the dietary intake of the element. The kidney must therefore be the organ principally responsible for regulating the total body content of magnesium. When dietary intake of magnesium is increased or decreased, urinary excretion of magnesium is increased or decreased respectively without any significant change in the plasma level of magnesium.

2. Effect of Dietary Restriction of Magnesium

Retention of magnesium by the kidney occurs rapidly in response to a restriction in the dietary intake (17, 43). This is why it is so difficult to produce magnesium depletion in the adult without some source of abnormal loss from the body.

Diurnal variations in the urinary excretion of calcium and magnesium have been demonstrated in patients in a metabolism ward (25). A reduction in the excretion of calcium, magnesium, sodium, and creatinine occurs at night. There are slight but constant diurnal variations in the serum concentration of calcium and magnesium with the values being lower in the morning than in the evening. Diet and physical activity appear to play the dominant roles in this diurnal fluctuation, but there also might be an associated rhythmicity in the function of the parathyroid gland.

3. Mechanism of Renal Excretion

The mechanism of excretion of magnesium by the mammalian kidney is still unclear (92). It could involve glomerular filtration and partial reabsorption of the filtered material by the renal tubules, or the filtered material could be completely reabsorbed and the excreted magnesium appear by tubular secretion, as is believed to occur with potassium. Tubular secretion of magnesium undoubtedly occurs in the aglomerular fish (20), but stop-flow studies with radioactive magnesium in dogs have produced conflicting evidence about secretion of magnesium by the tubules (50, 104). In the rabbit, the renal excretion of magnesium appears to be essentially glomerular; the tubular wall appears to be impermeable to magnesium throughout its length (109).

4. A Possible Renal Threshold

The amount of magnesium that is filtered at the glomerulus in an adult human is about 9.6 mEq/h, assuming a glomerular filtration rate of 130 ml/min, a total plasma magnesium concentration of 1.6 mEq/l and an ultrafiltrable fraction comprising 75% of the total. The mean rate of magnesium excretion in the urine (about 0.33 mEq/h) therefore represents only 3.5% of the filtered load. Moreover, the whole range of excretion observed under physiologic conditions in man can be explained if the tubular reabsorption of magnesium varies between 91 and 99% of the amount filtered at the glomerulus. In the rat (14), sheep (138), and cattle (122), there is evidence for the existence of a renal threshold for excretion of magnesium at a value close to the lower limit of the normal blood level. There is reduction in net tubular reabsorption of magnesium above a total serum magnesium concentration of 1.2–1.4 mEq/l; this could be due to either a decrease in the maximum capacity for tubular reabsorption or an increase in tubular secretion of magnesium.

5. Tubular Secretion

The possibility of secretion of magnesium by the renal tubules has been investigated under conditions of magnesium loading (61). At serum concentrations above 6.2 mEq/l, the amount excreted exceeded twice the filtered load, thus demonstrating tubular secretion of magnesium beyond any likely experimental error. The response to the administration of 2,4-dinitrophenol suggested that magnesium is also secreted by the tubules under physiologic conditions.

All the available evidence in the rat until recently have been consistent with a mechanism for magnesium excretion which involves reabsorption of the filtered material, with the excreted magnesium derived chiefly by tubular secretion. This secretion only appears to commence when the magnesium concentration in serum exceeds a critical value which is close to the lower limit of the normal range. However, studies with the stop-flow techniques did not find magnesium secretion in acutely magnesium-loaded rats undergoing mannitol or sulfate diuresis (9).

In the dog (93), magnesium excretion, like sodium and calcium excretion, is determined by filtration and reabsorption alone without evidence for tubular secretion. There is a maximal tubular reabsorptive capacity (Tm) for magnesium of approximately 11.5 μEq/min/kg body weight. The parathyroid hormone may directly enhance tubular reabsorption of magnesium.

E. Homeostasis

We do not understand yet the physiologic mechanisms which are responsible for maintaining the plasma magnesium concentration at a constant level (84). Both calcitonin (81) and parathormone may be involved. Nevertheless, animals and human beings on an adequate intake of magnesium do remain in magnesium balance, and the two chief regulatory sites appear to be the gastrointestinal tract and the kidney.

1. Effects of Parathyroid Hormone

There is considerable evidence for the hypothesis that the parathyroid hormone may help to control the concentration of plasma magnesium through a negative feedback mechanism (49, 60, 85).

Magnesium deficiency in the intact rat is accompanied by hypercalcemia and hypophosphatemia, provided the parathyroid glands are intact. The concentration of ionic calcium in plasma is elevated. In the absence of the parathyroid gland, magnesium-deficient rats do not develop hypercalcemia or hypophosphatemia. Moreover, parathyroidectomized animals with magnesium deficiency develop a concentration of ionized calcium in plasma that is lower than that observed in parathyroidectomized rats on a normal diet (52, 111).

These observations help to establish a relationship between an apparent increased function of the parathyroid gland and magnesium deficiency (111). Recent studies suggest that magnesium depletion may result in impaired synthesis or release of parathyroid hormone in man, or both (13, 31, 123). Parathyroid hormone responsiveness in hypomagnesemic patients may, at least in part, be dependent upon the adequacy of intracellular magnesium stores (101).

2. Effects of Hypermagnesemia

If parathyroid regulation is influenced by the concentration of magnesium in plasma, hypermagnesemia should diminish parathyroid gland activity (12). This hypothesis was tested in intact and chronically parathyroidectomized rats which were nephrectomized to eliminate the urinary excretion of calcium as a variable in the study. Isotonic magnesium chloride was administered subcutaneously to the experimental animals and normal saline was administered to the controls. A significant decrease in the concentration of ionic calcium was observed in the magnesium-treated animals with the intact parathyroid glands. In contrast, magnesium-treated parathyroidectomized animals failed to develop a significant change in the concentration of ionic calcium in comparison to saline-treated parathyroidectomized controls. These observations suggest that hypermagnesemia may inhibit parathyroid gland activity. The results are consistent with the hypothesis that the parathyroid regulatory mechanism which is involved in calcium homeostatis is modified by alterations in the concentration of plasma magnesium (51).

3. Perfusion Studies

The influence of the plasma magnesium concentration on parathyroid gland function was evaluated in goats and in a sheep by perfusion of the isolated parathyroid gland with whole blood of varying magnesium concentration (26). The concentration of parathyroid hormone in venous plasma from the gland was estimated by a specific radioimmunoassay. In each animal, the concentration of parathyroid hormone in the effluent plasma *diminished* when the concentration of magnesium was raised; the concentration of hormone *increased* when the concentration of magnesium was lowered. The response of the parathyroid hormone concentration to changes in plasma magnesium concentration occurred rapidly within minutes. Magnesium appeared to have a specific influence on the rate of release of parathyroid hormone.

4. Studies in Organ Culture

Sherwood et al. (113) recently developed an organ culture system utilizing normal bovine parathyroid tissue. Studies with this system provide direct evidence that the release of parathyroid hormone is inversely proportional to both the calcium and the magnesium ion concentrations. These two cations are equipotent in blocking hormone release.

5. Relationship between Bone and Extracellular Magnesium

Magnesium deficiency in the rat has been shown repeatedly to cause lowering of the magnesium concentration in bone (91). The observation of a close direct relationship between the magnesium concentration in the plasma and the femur of magnesium-deficient rats, calves, and man (10) supports the view that the skeleton provides the magnesium reserve in the body and suggests that there exists an equilibrium between the magnesium of the plasma and the bone. Recent clinical studies indicate that bone and extracellular fluid magnesium and the major magnesium pools in man increased during magnesium excess and decreased during magnesium depletion (11). This equilibrium is apparently independent of enzymatic activity and must, therefore, be physicochemical in nature. The fact that the equilibrium is dependent upon the concentration of magnesium in both the medium and the bone suggests that the relationship between bone and extracellular fluid magnesium is analogous to the ionization of a poorly dissociated salt, with the magnesium in bone corresponding to the undissociated salt.

6. Effects of Parathyroid Extract *in vitro*

Parathyroid extract increases the rate of magnesium loss from either fresh or boiled bone *in vitro* in a magnesium-low medium containing 50% bovine serum; however, the extract has no effect in a protein-free medium. These observations are consistent with the hypothesis that the physicochemical action of parathyroid preparations may involve the binding of divalent cations by a parathyroid-albumin complex (54, 91). This phenomenon in dead tissue, which may partially explain an important biologic function, certainly is not in accord with current concepts of the mechanism of hormonal action.

F. *Magnesium Deficiency in Man*

1. The Clinical Syndrome

For many years, there was doubt about the existence of a pure magnesium deficiency state in man. Now it is established that there is such a condition (44, 129). It is characterized by the following features: (1) spasmophilia (41), gross muscular tremor, choreiform movements, ataxia, tetany and, in some instances, predisposition to epileptiform convulsions (58); (2) hallucinations, agitation, confusion, tremulousness, delirium, depression, vertigo, muscular weakness, and an organic brain syndrome (57); (3) a low serum magnesium concentration associated with a normal serum calcium concentration and a normal blood pH; (4) a low-voltage T wave in the electrocardiogram (27), low-voltage PQRS complexes, and a short fixed P-R interval (16); (5) a positive Chvostek and Trousseau sign; and (6) prompt relief of the tetany when the serum magnesium concentration is restored to normal (129). *Durlach* (40) recognizes the presence of other manifestations of clinical magnesium deficiency, such as phlebothrombosis, constitutional thrombasthenia and hemolytic anemia, an allergic or osseous form of the deficiency, and oxalate lithiasis.

2. Experimental Production of a Pure Magnesium Deficiency

It is difficult to achieve a significant magnesium depletion in normal individuals by simple dietary restriction because of the exceedingly efficient renal and gastrointestinal mechanisms for conservation. The urinary magnesium in normal individuals falls to trivial amounts within 4—6 days of magnesium restriction (18, 43). In spite of these conservatory mechanisms, *Dunn and Walser* (39) did induce in two normal subjects deficits approaching 10% of the total body content of magnesium by infusing sodium sulfate and adding calcium supplements to the magnesium-deficient diet. The concentration of magnesium in plasma and erythrocytes fell moderately. Because the muscle magnesium content remained normal, the presumption was that bone was the source of the loss. No untoward clinical effects were noted.

Randall et al. (107) reported data suggesting that total body depletion of magnesium may result in psychiatric and neuromuscular symptoms. Administration of magnesium by the parenteral route or in the diet was associated with clinical improvement which occasionally was dramatic.

The best study to date of magnesium deficiency in man is that recently reported by *Shils* (114—116). Seven subjects were placed on a magnesium-deficient diet containing 0.7 mEq of magnesium/day. The concentration of magnesium in plasma declined perceptibly in all subjects within 7—10 days. Urinary and fecal magnesium decreased markedly, as did urinary calcium. At the height of the deficiency, the plasma magnesium concentration fell to a range of 10—30% of the control values, while the red cell magnesium declined more slowly and to a smaller degree. All male subjects developed hypocalcemia; the one female patient did not. Marked and persistent symptoms developed only in the presence of hypocalcemia. The serum potassium concentration decreased, and in four of the five subjects in whom the measurement was made, the ^{42}K space was decreased. The serum sodium concentration was not altered significantly. Three of the four subjects with the severest symptoms also had metabolic alkalosis.

A positive Trousseau sign which occurred in five of the seven subjects was the most common neurologic sign observed. Electromyographic changes, which were characterized by the development of myopathic potentials, occurred in all five of the patients tested. Anorexia, nausea, and vomiting were frequently experienced. When magnesium was added to the experimental diet, all clinical and biochemical abnormalities were corrected.

3. Clinical Conditions Associated with Depletion of Magnesium

Magnesium deficiency can occur in congestive heart failure, after diuresis, with furosemide, ethacrynic acid and mercurials, and with digitalis intoxication, diabetic acidosis, acute and chronic alcoholism, delirium tremens, cirrhosis, malabsorption syndromes, protracted postoperative cases, open heart surgery, the diuretic phase of acute tubular necrosis, and with primary hypoparathyroidism, primary aldosteronism, juxtaglomerular hyperplasia, and pancreatitis (68).

a) Fasting. Prolonged fasting is associated with a continued renal excretion of magnesium (38). After 2 months of fasting, the deficit in some subjects may amount to 20% of the total body content of magnesium. Despite evidence for depletion of magnesium in muscle, the concentration of magnesium in plasma remains unchanged. The excess acid load presented for excretion to the kidney and the absence of intake of carbohydrate might be factors contributing to the persistent loss of magnesium. The magnitude of the excretion of magnesium parallels the severity of the acidosis. The ingestion of glucose decreases the urinary loss of magnesium.

b) Excess loss from the gastrointestinal tract. Persistent vomiting or prolonged removal of intestinal secretions by mechanical suction coupled with the administration of magnesium-free intravenous infusions can induce clinical magnesium deficiency (48, 72).

c) Surgical patients. There are postoperative changes in magnesium metabolism in patients undergoing a variety of operations involving a moderate degree of trauma (59). A lowered serum magnesium concentration is observed on the day after operation in 56% of the patients, but it is usually corrected by the second or third postoperative day. Surgery is followed by a negative magnesium balance of days' duration and similar changes are observed after dietary restriction in normal subjects. However, the magnitude of the magnesium loss following surgery is minimal and usually does not result in symptomatic magnesium deficiency (73, 82, 103).

d) Gastrointestinal disorders. The intestinal tract plays a major role in magnesium homeostasis. The rate of transport of magnesium across the intestine appears to be slower than that of calcium and directly proportional to intestinal transit time (87). Malabsorption of magnesium, therefore, occurs in conditions in which intestinal transit is abnormally rapid or in which the major absorbing site, the distal small intestine, has been resected.

Hypomagnesemia is associated frequently with *malabsorption* due to a variety of causes. In general, there appears to be a correlation between the degree of hypomagnesemia and the severity of the underlying disease. The increased fecal loss of magnesium that has been demonstrated in this disorder may be due to steatorrhea (53). In acute pancreatitis, hypomagnesemia may occur due in part to deposition of this cation in areas of fat necrosis (63).

e) Acute alcoholism. The mean serum magnesium value in patients with delirium tremens in one study was 1.53 ± 0.27 mEq/l. In alcoholics without delirium tremens, it was 1.89 ± 0.22 mEq/l. In the control group of 157 non-alcoholics, the mean serum magnesium value was 1.84 ± 0.18 mEq/l. There was a tendency for the lowest serum magnesium levels to coincide with the highest values for serum glutamic oxalacetic transaminase (105). Hypomagnesemia occurs frequently in patients with chronic alcoholism with and without delirium tremens. Patients exhibiting alcohol withdrawal signs and symptoms (100) have low serum and cerebrospinal fluid levels of magnesium, low exchangeable magnesium levels (88, 99), a lowered muscle content of magnesium (70, 77), and conservation of magnesium following intravenous loading (96). A transient decrease in serum magnesium may occur during the withdrawal state even though prewithdrawal levels are normal. An ethanol-induced increase of magnesium in the urine occurs only when the blood alcohol level is rising. It does not persist once the subject has established high blood alcohol levels. However, in the presence of hypomagnesemia and delirium tremens, sudden death can occur as a result of cardiovascular collapse, infection, and hyperthemia (102). The red cell concentration of magnesium is abnormally low in all patients with delirium tremens, whereas the plasma concentration is abnormally low in only 58% of them (120). Intracellular fluid levels of magnesium as reflected in the erythrocyte correlate better with clinical symptoms and signs than do extracellular fluid levels. The predominant factor accounting for magnesium depletion in acute alcoholism is most likely an inadequate intake of magnesium, but another factor may be increased excretion of magnesium in the urine and feces (37, 97).

Independent of the phenomena described above, an abrupt and significant fall in serum magnesium levels may occur following cessation of drinking. This acute fall in serum magnesium level is associated with a transient decrease in concentration of other serum electrolytes and with respiratory alkalosis (139) and coincides with the onset of neuromuscular hyperexcitability that characterizes the withdrawal state (100). Hypomagnesemia appears to be directly related to the syndrome of alcoholic encephalopathy; adequate treatment with magnesium reverses the syndrome (121). A kinetic analysis of radiomagnesium turnover was performed in a group of partially repleted alcoholic subjects. Despite the continued presence of hypomagnesemia and of decreased urinary excretion of magnesium, there was little evidence of continued depletion of magnesium in the extracellular space or in the tissue pools (131).

f) Cirrhosis. The magnesium content of the liver tissue per unit weight is decreased in cirrhosis (136). This decrease appears to be due mainly to the substitution of parenchymal tissue of high magnesium content with connective tissue of low magnesium content. There is a good relationship between histological changes (extent of fibrosis and degree of infiltration of inflammatory cells) and decrease of the magnesium concentration per number of cells. The actual changes in the concentration of magnesium in the parenchymal cells of the cirrhotic liver appear to be negligible.

Patients with cirrhosis may have clinical features consistent with magnesium deficiency in the presence of a normal serum magnesium value but with a low skeletal muscle magnesium content and a normal bone and erythrocyte magnesium content (78).

g) Cardiovascular disorders. It was recognized but not widely appreciated as early as 1952 that magnesium deficiency occurs in congestive heart failure (67), and that hypomagnesemia follows the administration of mercuhydrin (90). Cardiac glycosides may induce magnesium deficiency; magnesium deficiency is frequently associated with cardiac arrhthymias such as ventricular tachycardia and arterial or ventricular fibrillation (68).

h) Hypomagnesemia as a cause of persistent hypokalemia. Of considerable clinical interest is the recent recognition that hypokalemia may be secondary to magnesium deficiency and may be resistant to treatment unless the underlying magnesium deficiency is corrected (133). Since hypokalemia is so prevalent, a high index of suspicion is necessary to detect the underlying magnesium deficiency.

IV. Diagnosis

In many patients, the clinical symptoms and signs, although non-specific, accompanied by a low serum magnesium concentration confirm the diagnosis. However, a normal serum magnesium level does not exclude magnesium deficiency.

Since serum magnesium is regulated largely by renal control of urinary magnesium secretion, urinary output of the element has been used as an index of magnesium deficiency (61). *Caddell* (28) has described a magnesium load test in infants up to 6 months of age. A 56-hour test measured cation and creatinine excretion before and after an intramuscular load of 0.49 mEq Mg/kg of body weight. This approach has been extended to postpartum American women (29). An intravenous test load of 0.4–0.5 mEq Mg/kg of estimated lean body weight was administered and the net retention calculated. *Thoren* (127) has used an intravenous dose of 0.25 mmol of Mg/kg of body weight; more than 80% of this dose should be excreted in the urine within 24 h if tissue reserves are adequate.

V. Therapy

In patients with the clinical symptoms and signs of magnesium deficiency, the deficit of magnesium is on the order of 1–2 mEq/kg of body weight. Since less than one half of the administered magnesium is usually retained in the body, the required therapeutic dose is 2—4 mEq/kg which can be administered parenterally over a 4-day period. In order to administer the dose safely, one should first determine the adequacy of renal function, then monitor the plasma levels of magnesium during therapy (45).

Flink (45) recommends, an initial loading dose of 49 mEq (6.0 g of $MgSO_4$ in 1,000 ml of solution containing 5% glucose) given intravenously over a period of 3 h, followed by additional doses of 49 mEq every 12 h.

Another suggested regimen is the intravenous administration of 98 mEq on the first day (2.0 g, 16.3 mEq, every 2 h for three doses and then every 4 h for four doses), followed by 33—49 mEq/day in divided doses for 4 days (45).

For the treatment of arrhythmia, *Iseri et al.* (68) administer 10—15 ml of a 20% magnesium sulfate solution, given intravenously over 1 min, followed by a slow 4- to 6-hour infusion of 500 ml of a 2% magnesium sulfate in 5% dextrose water. A second infusion of magnesium sulfate may be necessary should the arrhythmia recur.

VI. Summary and Comments

Almost seven decades have passed since *Richard Willstatter* demonstrated the central position of the magnesium atom in the chlorophyll molecule. Although much has been learned since concerning the photosynthetic process, the exact function of the magnesium atom in this process still eludes us. That magnesium is essential for photosynthesis is an established fact. There is recent evidence to suggest that magnesium may play a role in the regulation of the photosynthetic processes.

Chlorophyll is a component of the chloroplast which is located physically in the granum which possesses a lamellar structure conductive to charge separation; the granum is the locus for photosynthesis. What is the role of magnesium at this interface of chemistry and physics to molecular biology? What about the properties of water with its low-energy bonds as emphasized by *Szent-Györgi* (126)? Can polarized water be as important as lipid in providing the living cell with its selective surface barrier (80)? Can all of the functions of magnesium be explained solely on the basis of the coordinating properties of the magnesium atom?

In the synthetic processes which follow the capture of solar energy, ATP plays a vital role. It is significant that all enzyme reactions which are known to be catalyzed by ATP have an absolute requirement for magnesium; so fundamental and widespread are the reactions involving ATP that it must influence practically all processes of life.

Is the mitochondrion a single large branching tubular structure *in vivo* and are the small intracellular organelles artifacts of preparation; what is the role of magnesium in this living organelle? Recent attempts to use computer graphics and computer analysis of serial section electron micrographs may supply further insight into these problems which involve detailed studies of the tertiary structure and the spacial interrelationships of terribly complex molecules.

Energy originally derived from the sun is used by living organisms for genetic transcription and protein synthesis. It appears that magnesium plays a vital role in all of these processes: the physical integrity of the DNA helix is dependent upon magnesium; the physical size of the RNA aggregates is controlled by the concentration of magnesium; and polypeptide formation is magnesium-dependent. How to explain this all-pervasive role of magnesium?

Concerning the physiology of magnesium, the first obvious fact is that most of it is located within cells. This differential concentration is on the order of 10:1, intracellular/extracellular, in the soft tissues of the body. Is metabolic energy required for the maintenance of this state, or can this be explained as due primarily to physicochemical adsorption? Can the symptoms and signs of magnesium deficiency be explained simply on the basis of the coordinating properties of the magnesium atom?

Bone contains the highest concentration of magnesium of any tissue in the body. Does this concentration require the expenditure of metabolic energy? How exactly are the parathyroid hormone and calcitonin involved in this process?

These are questions begging further studies.

References

1   *Aikawa, J.K.:* Mg^{28} tracer studies of magnesium metabolism in animals and human beings. Proc. 2nd UN Int. Conf. Peaceful Uses Atomic Energy, Geneva, vol. 24, pp. 148–151 (Pergamon, London 1958).

2   *Aikawa, J.K.:* Gastrointestinal absorption of Mg^{28} in rabbits. Proc. Soc. exp. Biol. Med. *100:* 293–295 (1959).

3   *Aikawa, J.K.:* The role of magnesium in biologic processes. A review of recent developments; in *Bajusz* Electrolytes and cardiovascular diseases, pp. 9–27 (Karger, Basel 1965).

4   *Aikawa, J.K.:* The relationship of magnesium to disease in domestic animals and in humans (Thomas, Springfield 1971).

5   *Aikawa, J.K.; Gordon, G.S., and Rhoades, E.L.:* Magnesium metabolism in human beings: studies with Mg^{28}. J. appl. Physiol. *15:* 503–507 (1960).

6   *Aikawa, J.K.; Rhoades, E.L., and Gordon, G.S.:* Urinary and fecal excretion of orally administered Mg^{28}. Proc. Soc. exp. Biol. Med. *98:* 29–31 (1958).

7   *Aikawa, J.K.; Rhoades, E.L.; Harms, D.R., and Reardon, J.Z.:* Magnesium metabolism in rabbits using Mg^{28} as a tracer. Am. J. Physiol. *197:* 99–101 (1959).

8 *Alcock, N.W. and MacIntyre, I.:* Inter-relation of calcium and magnesium absorption. Clin. Sci. *22:* 185–193 (1962).

9 *Alfredson, K.S. and Walser, M.:* Is magnesium secreted by the rat renal tubule? Nephron *7:* 241–247 (1970).

10 *Alfrey, A.C. and Miller, N.L.:* Bone magnesium pools in uremia. J. clin. Invest. *52:* 3019–3027 (1973).

11 *Alfrey, A.C.; Miller, N.L., and Butkus, D.:* Evaluation of body magnesium stores. J. Lab. clin. Med. *84:* 153–162 (1974).

12 *Altenähr, E. and Leonhardt, F.:* Suppression of parathyroid gland activity by magnesium. Virchows Arch. Abt. A Path. Anat. *355:* 297–308 (1972).

13 *Anast, C.S.; Mohs, J.M.; Kaplan, S.L., and Burns, T.W.:* Evidence for parathyroid failure in magnesium deficiency. Science *177:* 606–608 (1972).

14 *Averill, C.M. and Heaton, F.W.:* The renal handling of magnesium. Clin. Sci. *31:* 353–360 (1966).

15 *Avioli, L.V. and Berman, M.:* Mg^{28} kinetics in man. J. appl. Physiol. *21:* 1688–1694 (1966).

16 *Bajpai, P.C.; Hasan, M.; Gupta, A.K., and Kanwar, K.B.:* Electrocardiographic changes in hypomagnesemia. Indian Heart J. *24:* 271–276 (1972).

17 *Barnes, B.A.; Cope, O., and Harrison, T.:* Magnesium conservation in the human being on low magnesium diet. J. clin. Invest. *37:* 430–440 (1958).

18 *Barnes, B.A.; Cope, O., and Gordon, E.B.:* Magnesium requirements and deficits. An evaluation of two surgical patients. Ann. Surg. *152:* 518–533 (1960).

19 *Baron, D.N. and Ahmet, S.A.:* Intracellular concentrations of water and of the principal electrolytes determined by analysis of isolated human leucocytes. Clin. Sci. *37:* 205–219 (1969).

20 *Berglund, F. and Forster, R.P.:* Renal tubular transport of inorganic divalent ions by the aglomerular teleost, *Lophius americanus.* J. gen. Physiol. *41:* 429–440 (1958).

21 *Binet, A. and Volfin, P.:* Regulation by Mg^{2+} and Ca^{2+} of mitochondrial membrane integrity. Study of the effects of a cytosolic molecule and Ca^{2+} antagonists. Archs Biochem. *170:* 576–586 (1975).

22 *Bogucka, K. and Wojtczak, L.:* Binding of magnesium by proteins of the mitochondrial intermembrane compartment. Biochem. biophys. Res. Commun. *71:* 161–167 (1976).

23 *Brandt, J.T.; Martin, A.P.; Lucas, F.V., and Vorbeck, M.L.:* The structure of normal rat liver mitochondria: a re-evaluation. Fed. Proc. Fed. Am. Socs exp. Biol. *33:* 603 (1974).

24 *Breibart, S.; Lee, J.S.; McCoord, A., and Forbes, G.:* Relation of age to radiomagnesium in bone. Proc. Soc. exp. Biol. Med. *105:* 361–363 (1960).

25 *Briscoe, A.M. and Ragan, C.:* Diurnal variations in calcium and magnesium excretion in man. Metabolism *15:* 1002–1010 (1966).

26 *Buckle, R.M.; Care, A.D.; Cooper, C.W., and Gitelman, H.J.:* The influence of plasma magnesium concentration on parathyroid hormone secretion. J. Endocr. *42:* 529–534 (1968).

27 *Caddell, J.L.:* Studies in protein-calorie malnutrition. II. A double-blind clinical trial to assess magnesium therapy. New Engl. J. Med. *276:* 535–540 (1967).

28 *Caddell, J.L.:* The magnesium load test. I. A design for infant. Clin. Pediat. *14:* 449–459 (1975).

29 *Caddell, J.L.; Saier, F.L., and Thomason, C.A.:* Parenteral magnesium load tests in postpartum American women. Am. J. clin. Nutr. *28:* 1099–1104 (1975).

30 *Case, G.S.; Sinnott, M.L., and Tenu, J.-P.:* The role of magnesium ions in β-galactosidase-catalyzed hydrolyses. Biochem. J. *133:* 99–104 (1973).

31 *Chase, L.R. and Slatopolsky, E.:* Secretion and metabolic efficacy of parathyroid hormone in patients with severe hypomagnesemia. J. clin. Endocr. Metab. *38:* 363–371 (1974).

32 *Chesley, L.C. and Tepper, I.:* Some effects of magnesium loading upon renal excretion of magnesium and certain other electrolytes. J. clin. Invest. *37:* 1362–1372 (1958).

33 *Chutkow, J.G.:* Sites of magnesium absorption and excretion in the intestinal tract of the rat. J. Lab. clin. Med. *63:* 71–79 (1964).

34 *Clement, R.M.; Sturm, J., and Daune, M.D.:* Interaction of metallic cations with DNA. VI. Specific binding of Mg^{++} and Mn^{++}. Biopolymers *12:* 405–421 (1973).

35 *Consolazio, C.F.; Matoush, L.O.; Nelson, R.A.; Harding, R.S., and Canham, J.E.:* Excretion of sodium, potassium, magnesium and iron in human sweat and the relation of each to balance and requirements. J. Nutr. *79:* 407–415 (1963).

36 *Coussons, H.:* Magnesium metabolism in infants and children. Postgrad. Med. *46:* 135–139 (1969).

37 *Dick, M.; Evans, R.A., and Watson, L.:* Effect of ethanol on magnesium excretion. J. clin. Path. *22:* 152–153 (1969).

38 *Drenick, E.J.; Hunt, I.F., and Swendseid, M.E.:* Magnesium depletion during prolonged fasting of obese males. J. clin. Endocr. Metab. *29:* 1341–1348 (1969).

39 *Dunn, M.J. and Walser, M.:* Magnesium depletion in normal man. Metabolism *15:* 884–895 (1966).

40 *Durlach, J.:* Le magnésium en pathologie humaine. Problèmes pratiques et incidence diététiques. Gaz. méd. Fr. *74:* 3303–3320 (1967).

41 *Durlach, J.; Gremy, F. et Metral, S.:* La spasmophilie: forme clinique neuro-musculaire du déficit magnésien primitif. Revue neurol. *117:* 177–189 (1967).

42 *Field, A.C.:* Magnesium in ruminant nutrition. III. Distribution of Mg^{28} in the gastrointestinal tract and tissues of sheep. Br. J. Nutr. *15:* 349–359 (1961).

43 *Fitzgerald, M.G. and Fourman, P.:* An experimental study of magnesium deficiency in man. Clin. Sci. *15:* 635–647 (1956).

44 *Flink, E.B.:* Magnesium deficiency syndrome in man. J. Am. med. Ass. *160:* 1406–1409 (1956).

45 *Flink, E.B.:* Therapy of magnesium deficiency. Ann. N.Y. Acad. Sci. *162:* 901–905 (1969).

46 *Forster, R.P. and Berglund, F.:* Osmotic diuresis and its effect on total electrolyte distribution in plasma and urine of the aglomerular teleost, *Lophius americanus.* J. gen. Physiol. *39:* 349–359 (1956).

47 *Fuhrhop, J.-H. and Mauzerall, D.:* The one-electron oxidation of metalloporphyrins. J. Am. chem. Soc. *91:* 4174–4181 (1969).

48 *Gerst, P.H.; Porter, M.R., and Fishman, R.A.:* Symptomatic magnesium deficiency in surgical patients. Ann. Surg. *159:* 402–406 (1964).

49 *Gill, J.R., jr.; Bell, N.H., and Bartter, F.C.:* Effect of parathyroid extract on magnesium excretion in man. J. appl. Physiol. *22:* 136–138 (1967).

50 *Ginn, H.E.; Smith, W.O.; Hammarsten, J.R., and Snyder, D.:* Renal tubular secretion of magnesium in dogs. Proc. Soc. exp. Biol. Med. *101:* 691–692 (1959).

51 *Gitelman, H.J.; Kukolj, S., and Welt, L.G.:* Inhibition of parathyroid gland activity by hypermagnesemia. Am. J. Physiol. *215:* 483–485 (1968).

52 *Gitelman, H.J.; Kukolj, S., and Welt, L.G.:* The influence of the parathyroid glands on the hypercalcemia of experimental magnesium depletion in the rat. J. clin. Invest. *47:* 118–126 (1968).

53 *Gitelman, J.H. and Welt, L.G.:* Magnesium deficiency. Annu. Rev. Med. *20:* 233–242 (1969).

54 *Gordon, G.S.:* A direct action of parathyroid hormone on dead bone *in vitro.* Acta endocr., Copenh. *44:* 481–489 (1963).

55 *Graham, L.A.; Caesar, J.J., and Burgen, A.S.V.:* Gastrointestinal absorption and excretion of Mg²⁸ in man. Metabolism *9:* 646–659 (1960).

56 *Green, D.E.:* The mitochondrion. Sci. Am. *210:* 63–74 (1964).

57 *Hall, R.C.W. and Jaffe, J.R.:* Hypomagnesemia. Physical and psychiatric symptoms. J. Am. med. Ass. *224:* 1749–1751 (1973).

58 *Hanna, S.; Harrison, M.; MacIntyre, I., and Fraser, R.:* The syndrome of magnesium deficiency in man. Lancet *ii:* 172–176 (1960).

59 *Heaton, F.W.:* Magnesium metabolism in surgical patients. Clin. chim. Acta *9:* 327–333 (1964).

60 *Heaton, F.W.:* The parathyroid and magnesium metabolism in the rat. Clin. Sci. *28:* 543–553 (1965).

61 *Heaton, F.W.:* The kidney and magnesium homeostasis. Ann. N.Y. Acad. Sci. *162:* 775–785 (1969).

62 *Hendrix, J.Z.; Alcock, N.W., and Archibald, R.M.:* Competition between calcium, strontium, and magnesium for absorption in the isolated rat intestine. Clin. Chem. *9:* 734–744 (1963).

63 *Hersh, T. and Siddiqui, D.A.:* Magnesium and the pancreas. Am. J. clin. Nutr. *26:* 362–366 (1973).

64 *Hirschfelder, A.D. and Haury, V.G.:* Clinical manifestations of high and low plasma magnesium; dangers of epsom salt purgation in nephritis. J. Am. med. Ass. *102:* 1138–1141 (1934).

65 *Hoffman, H.-P. and Avers, C.J.:* Mitochondrion of yeast. Ultrastructural evidence for one giant, branched organelle per cell. Science *181:* 749–751 (1973).

66 *Hughes, M.N.:* The inorganic chemistry of biological processes (Wiley, Chichester 1972).

67 *Iseri, L.T.; Alexander, L.C.; McCaughey, R.S.; Boyle, A.J., and Myers, G.B.:* Water and electrolyte content of cardiac and skeletal muscle in heart-failure and myocardial infarction. Am. Heart J. *43:* 215–227 (1952).

68 *Iseri, L.T.; Freed, J., and Bures, A.R.:* Magnesium deficiency and cardiac disorders. Am. J. Med. *58:* 837–846 (1975).

69 *Jones, J.E.; Manalo, R., and Flink, E.B.:* Magnesium requirements in adults. Am. J. clin. Nutr. *20:* 632–635 (1967).

70 *Jones, J.E.; Shane, S.R.; Jacobs, W.H., and Flink, E.B.:* Magnesium balance studies in chronic alcoholism. Ann. N.Y. Acad. Sci. *162:* 934–946 (1969).

71 *Keating, F.R.; Jones, J.D.; Elveback, L.R., and Randall, R.V.:* The relation of age and sex to distribution of values in healthy adults of serum calcium, inorganic phosphorus, magnesium, alkaline phosphatase, total proteins, albumin, and blood urea. J. Lab. clin. Med. *73:* 825–834 (1969).

72 *Kellaway, G. and Ewen, K.:* Magnesium deficiency complicating prolonged gastric suction. N.Z. med. J. *6:* 137–142 (1962).

73 *King, L.R.; Knowles, H.C., jr., and McLaurin, R.I.:* Calcium, phosphorus, and magnesium metabolism following head injury. Ann. Surg. *177:* 126–131 (1963).

74 *Krakauer, H.:* A calorimetric investigation of the heats of binding of Mg⁺⁺ to Poly A, to Poly U, and to their complexes. Biopolymers *11:* 811–828 (1972).

75 *Lear, R.D. and Grøn, P.:* Magnesium in human saliva. Archs oral Biol. *13:* 1311–1319 (1968).

76 *Lengemann, F.W.:* The metabolism of magnesium and calcium by the rat. Archs Biochem. *84:* 278–285 (1959).

77 Lim, P. and Jacob, E.: Magnesium status of alcoholic patients. Metabolism *21:* 1045–1051 (1972).

78 Lim, P. and Jacob, E.: Magnesium deficiency in liver cirrhosis. Q. Jl Med. *41:* 291–300 (1972).

79 Lin, D.C. and Nobel, P.S.: Control of photosynthesis by Mg^{2+}. Archs Biochem. *145:* 622–632 (1971).

80 Ling, G.W.: What component of the living cell is responsible for its semipermeable properties? Polarized water or lipid? Biophys. J. *13:* 807–816 (1973).

81 Littledike, E.T. and Arnaud, C.D.: The influence of plasma magnesium concentrations on calcitonin secretion in the pig. Proc. Soc. exp. Biol. Med. *136:* 1000–1006 (1971).

82 Macbeth, R.A.L. and Mabbott, J.D.: Magnesium balance in the postoperative patient. Surgery Gynec. Obstet. *118:* 748–760 (1964).

83 MacIntyre, I.: Discussion on magnesium metabolism in man and animals. Proc. R. Soc. Med. *53:* 1037–1039 (1960).

84 MacIntyre, I.: Magnesium metabolism. Adv. intern. Med. *13:* 143–154 (1967).

85 MacIntyre, I.; Boss, S., and Troughton, V.A.: Parathyroid hormone and magnesium homeostasis. Nature, Lond. *198:* 1058–1060 (1963).

86 MacIntyre, I.; Hanna, S.; Booth, C.C., and Read, A.E.: Intracellular magnesium deficiency in man. Clin. Sci. *20:* 297–305 (1961).

87 MacIntyre, I. and Robinson, C.J.: Magnesium in the gut: experimental and clinical observations. Ann. N.Y. Acad. Sci. *162:* 865–873 (1969).

88 Martin, H.E. and Bauer, F.K.: Magnesium 28 studies in the cirrhotic and alcoholic. Proc. R. Soc. Med. *55:* 912–914 (1962).

89 Martin, H.E.; McCuskey, C., and Tupikova, N.: Electrolyte disturbance in acute alcoholism. With particular reference to magnesium. Am. J. clin. Nutr. *7:* 191–196 (1959).

90 Martin, H.E.; Mehl, J., and Wertman, M.: Clinical studies of magnesium metabolism. Med. Clins N. Am. *36:* 1157–1171 (1952).

91 Martindale, L. and Heaton, F.W.: The relation between skeletal and extra-cellular fluid *in vitro.* Biochem. J. *97:* 440–443 (1965).

92 Massry, S.G. and Coburn, J.W.: The hormonal and non-hormonal control of renal excretion of calcium and magnesium. Nephron *10:* 66–112 (1973).

93 Massry, S.G.; Coburn, J.W., and Kleeman, C.R.: Renal handling of magnesium in the dog. Am. J. Physiol. *216:* 1460–1467 (1969).

94 Mauzerall, D.: Why chlorophyll? Ann. N.Y. Acad. Sci. *206:* 483–494 (1973).

95 McAleese, E.M.; Bell, M.C., and Forbes, R.M.: Mg^{28} studies in lambs. J. Nutr. *74:* 505–514 (1960).

96 McCollister, R.J.; Flink, E.B., and Doe, R.P.: Magnesium balance studies in chronic alcoholism. J. Lab. clin. Med. *55:* 98–104 (1960).

97 McCollister, R.J.; Flink, E.B., and Lewis, M.D.: Urinary excretion of magnesium in man following the ingestion of ethanol. Am. J. clin. Nutr. *12:* 415–420 (1963).

98 Mendel, L.B. and Benedict, S.R.: The paths of excretion for inorganic compounds. IV. The excretion of magnesium. Am. J. Physiol. *25:* 1–22 (1909).

99 Mendelson, J.H.; Barnes, B.; Mayman, C., and Victor, M.: The determination of exchangeable magnesium in alcoholic patients. Metabolism *14:* 88–98 (1965).

100 Mendelson, J.H.; Ogata, M., and Mello, N.K.: Effects of alcohol ingestion and withdrawal on magnesium states of alcoholics: clinical and experimental findings. Ann. N.Y. Acad. Sci. *162:* 918–933 (1969).

101 Michelis, M.F.; Bragdon, R.W.; Fusco, F.D.; Eichenholz, A., and David, B.B.: Parathyroid hormone responsiveness in hypoparathyroidism with hypomagnesemia. Am. J. med. Sci. *270:* 412–418 (1975).

102 *Milner, G. and Johnson, J.:* Hypomagnesaemia and delirium tremens. Report of a case with fatal outcome. Am. J. Psychiat. *122:* 701–702 (1965).

103 *Monsaingeon, A.; Thomas, J.; Nocquet, Y.; Savel, J. et Clostre, F.:* Sur le rôle du magnésium en pathologie chirurgicale. J. Chir., Paris *91:* 437–454 (1966).

104 *Murdaugh, H.V. and Robinson, R.R.:* Magnesium excretion in the dog studied by stop-flow analysis. Am. J. Physiol. *198:* 571–574 (1960).

105 *Nielsen, J.:* Magnesium metabolism in acute alcoholics. Dan. med. Bull. *10:* 225–233 (1963).

106 *Racker, E.:* Inner mitochondrial membranes: basic and applied aspects. Hosp. Pract. *9:* 87–93 (1974).

107 *Randall, R.E.; Rossmeisl, E.C., and Bleifer, K.H.:* Magnesium depletion in man. Ann. intern. Med. *50:* 257–287 (1959).

108 *Rasmussen, H.:* Ions as 'second messengers'. Hosp. Pract. *9:* 99–107 (1974).

109 *Raynaud, C.:* Renal excretion of magnesium in the rabbit. Am. J. Physiol. *203:* 649–654 (1962).

110 *Rubin, H.:* Central role for magnesium in coordinate control of metabolism and growth in animal cells. Proc. natn. Acad. Sci. USA *72:* 3551–3555 (1975).

111 *Sallis, J.D. and DeLuca, H.F.:* Action of parathyroid hormone on mitochondria. Magnesium- and phosphate-independent respiration. J. biol. Chem. *241:* 1122–1127 (1966).

112 *Schroeder, H.A.; Nason, A.P., and Tipton, I.H.:* Essential metals in man: magnesium. J. chron. Dis. *21:* 815–841 (1969).

113 *Sherwood, L.M.: Herrman, I., and Bassett, C.A.:* Parathyroid hormone secretion *in vitro.* Regulation by calcium and magnesium ions. Nature, Lond. *225:* 1056–1057 (1970).

114 *Shils, M.E.:* Experimental human magnesium depletion. I. Clinical observations and blood chemistry alterations. Am. J. clin. Nutr. *15:* 133–143 (1964).

115 *Shils, M.E.:* Experimental production of magnesium deficiency in man. Ann. N.Y. Acad. Sci. *162:* 847–855 (1969).

116 *Shils, M.E.:* Experimental human magnesium depletion. Medicine, Baltimore *48:* 61–85 (1969).

117 *Silver, L.; Robertson, J.S., and Dahl, L.K.:* Magnesium turnover in the human studied with Mg^{28}. J. clin. Invest. *39:* 420–425 (1960).

118 *Smith, P.K.; Winkler, A.W., and Schwartz, B.M.:* The distribution of magnesium following the parenteral administration of magnesium sulfate. J. biol. Chem. *129:* 51–56 (1939).

119 *Smith, R.H. and McAllan, A.B.:* Binding of magnesium and calcium in the contents of the small intestine in the calf. Br. J. Nutr. *20:* 703–718 (1966).

120 *Smith, W.O. and Hammarsten, J.F.:* Intracellular magnesium in delirium tremens and uremia. Am. J. med. Sci. *237:* 413–417 (1959).

121 *Stendig-Lindberg, G.:* Hypomagnesemia in alcohol encephalopathies. Acta psychiat. scand. *50:* 465–480 (1974).

122 *Storry, J.E. and Rook, J.A.F.:* The magnesium nutrition of the dairy cow in relation to the development of hypomagnesaemia in the grazing animal. J. Sci. Food Agric. *13:* 621–627 (1962).

123 *Suh, S.M.; Trashjian, A.H., jr.; Matsuo, N.; Parkinson, D.K., and Fraser, D.:* Pathogenesis of hypocalcemia in primary hypomagnesemia. Normal end-organ responsiveness to parathyroid hormone, impaired parathyroid gland function. J. clin. Invest. *52:* 153–160 (1973).

124 *Szent-Györgi, A.:* Introduction to a submolecular biology (Academic Press, New York 1960).

125 *Szent-Györgi, A.:* The ATP molecule; in *Kalckar* Biological phosphorylations. Development of concepts, pp. 486–493 (Prentice-Hall, Englewood Cliffs 1969).

126 *Szent-Györgi, A.:* Biology and pathology of water. Perspect. Biol. Med. *14:* 239–249 (1971).

127 *Thoren, L.:* Magnesium metabolism. Prog. Surg., vol. 9, pp. 131–156 (Karger, Basel 1971).

128 *Tibbetts, D.M. and Aub, J.C.:* Magnesium metabolism in health and disease. I. The magnesium and calcium excretion of normal individuals, also the effects of magnesium, chloride and phosphate ions. J. clin. Invest. *16:* 491–501 (1937).

129 *Wacker, W.E.C.; Moore, F.D.; Ulmer, D.D., and Vallee, B.L.:* Normocalcemic magnesium deficiency tetany. J. Am. med. Ass. *180:* 161–163 (1962).

130 *Wacker, W.E.C. and Parisi, A.F.:* Magnesium metabolism. New Engl. J. Med. *278:* 658–662, 712–717, 772–776 (1968).

131 *Wallach, S. and Dimich, A.:* Radiomagnesium turnover studies in hypomagnesemic states. Ann. N.Y. Acad. Sci. *162:* 963–972 (1969).

132 *Wallach, S.; Rizek, J.E.; Dimich, A.; Prasad, N., and Siler, W.:* Magnesium transport in normal and uremic patients. J. clin. Endocr. Metab. *26:* 1069–1080 (1966).

133 *Webb, S. and Schade, D.S.:* Hypomagnesemia as a cause of persistent hypokalemia. J. Am. med. Ass. *233:* 23–24 (1975).

134 *Wehrle, J.P.; Jurkowitz, M.; Scott, K.M., and Brierley, G.P.:* Mg^{28} and the permeability of heart mitochondria to monovalent cations. Archs Biochem. *174:* 312–323 (1976).

135 *Widdowson, E.M.; McCance, R.A., and Spray, C.M.:* The chemical composition of the human body. Clin. Sci. *10:* 113–125 (1951).

136 *Wilke, H. und Spielmann, H.:* Untersuchungen über den Magnesiumgehalt der Leber bei der Cirrhose. Klin. Wschr. *46:* 1162–1164 (1968).

137 *Willick, G.E. and Kay, C.M.:* Magnesium-induced conformational change in transfer ribonucleic acid as measured by circular dichroism. Biochemistry *10:* 2216–2222 (1971).

138 *Wilson, A.A.:* Magnesium homeostasis and hypomagnesaemia in ruminants. Vet. Rev. *6:* 39–52 (1960).

139 *Wolfe, S.M. and Victor, M.:* The relationship of hypomagnesemia and alkalosis to alcohol withdrawal symptoms. Ann. N.Y. Acad. Sci. *162:* 973–984 (1969).

140 *Zitomer, R.S. and Flaks, J.G.:* Magnesium dependence and equilibrium of the *Escherichia coli* ribosomal subunit association. J. molec. Biol. *71:* 263–279 (1972).

141 *Zumoff, B.; Bernstein, E.H.; Imarisio, J.J., and Hellman, L.:* Radioactive magnesium (Mg^{28}) metabolism in man. Clin. Res. *6:* 260 (1958).

J.K. Aikawa, MD, Professor of Medicine, University of Colorado School of Medicine, *Denver, CO 80262* (USA)

Wld Rev. Nutr. Diet., vol. 28, pp. 143–187 (Karger, Basel 1978)

Bone Growth and Development in Protein-Calorie Malnutrition

John H. Himes

Section of Physical Growth and Genetics, Fels Research Institute, Yellow Springs, Ohio

Contents

I. Introduction

Much attention regarding nutritional aspects of skeletal growth and development has been given studies documenting alterations in bone associated with deficiencies in vitamins and minerals. This concern is understandable in light of the established roles in bone growth and mineralization of vitamins, e.g. C and D, and minerals, notably calcium and phosphorus. Nevertheless, relatively little attention has been given to the effects on bone of general deficiency of protein and calories. This lack of emphasis is, in a sense, somewhat surprising in light of the prominent position that protein-calorie malnutrition (PCM) holds among world nutrition problems.

The World Health Organization (WHO, 1972) has presented those nutritional diseases that are most important on a global basis, and deserve most immediate attention; PCM is at the top of this list. Reliable estimates of the occurrence of PCM world-wide are difficult to obtain. *Bengoa* (1974), however, conservatively estimated some 100 million children, less than 4 years of age, are malnourished in Latin America, Africa, and Asia (excluding China and Japan). In 1968, *Béhar* made what he considered to be a conservative estimate that 300 million children, world-wide, less than 5 years old, suffered from at least mild to moderate forms of PCM; this amounted to about 60% of the total preschool-age population of the world at that time. By either estimate, it is clear that PCM, particularly in young children, is an overwhelming world problem. The prevalence of PCM and many of its consequences have been discussed in detail elsewhere (*Trowell et al.,* 1954; *Béhar,* 1968; *Jelliffe,* 1966; *Waterlow and Alleyne,* 1971; *Bengoa,* 1974; *Himes,* in press).

The purpose of the present discussion is to review the effects of PCM on various aspects of bone growth and development. There is an enormous body of literature dealing with somatic growth and PCM. Although external body dimensions, such as stature in man, or limb lengths in animals, are largely a function of elongation of long bones and vertebrae, the present discussion will deal only with studies concerning skeletons or bones *per se,* or their images on radiographs. Further, this review considers only associations of PCM and bone during the period of growth and development, and does not include studies of effects of PCM on bone in adults. It is not within the purview of the present discussion to give a detailed account of normal bone physiology, growth, composition, and maturation; rather, the reader is referred to a number of excellent comprehensive reviews on these subjects (*Acheson,* 1966; *Widdowson and Dickerson,* 1964; *Vaughan,* 1970; *Bourne,* 1971; *Ham,* 1974; *Roche et al.,* 1975; *Roche,* in press).

PCM in the present discussion includes deficiencies in protein and/or calories. PCM may be produced in experimental animals by starvation; or dietary restrictions in proteins and/or calories, or specific essential amino acids. PCM is more difficult to identify in human populations, often being complicated by disease or other nutrient deficiencies. Many investigators allude to PCM in discussions of results, or in defense of hypotheses, but present no evidence that the subjects are, in fact, malnourished. In order to reduce subjectivity, in the present review only those studies of humans that contain reasonable documentation of protein-calorie deficiency, or skeletal response to protein or calorie supplementation, have been included. Consequently, some studies have been excluded in which PCM may have been a factor. Similarly, many data regarding bone development of human infants considered small-for-date have not been included because the precise etiology of low birth weight is not completely known. Studies concerning bone growth and development of infants from mothers deprived of, or supplemented with protein and calories during pregnancy, however, are considered. Finally, several studies have examined the effects of milk in the diet on various aspects of skeletal growth and development. These have not been relied on heavily because the bone responses to milk could arise from calcium as well as protein and/or calories.

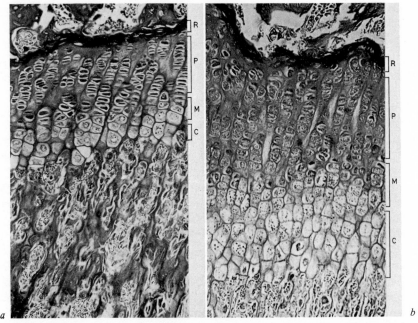

Fig. 1. Tibial epiphyseal plates of *(a)* rat fasted for 2 days, and *(b)* well-fed control. Bone growth zones: R = resting cartilage; P = proliferating cartilage; M = maturing cartilage; C = calcifying cartilage. Paraffin. × 400. Reprinted with permission from *Dearden and Mosier* (1974).

II. Histological Aspects

With PCM, endochondral bone growth is impaired. Perhaps the most frequently reported histological findings in this regard are associated with narrowed epiphyseal cartilages compared to well-fed controls. The extent of the narrowing of the epiphyseal cartilage appears related to the degree and duration of the undernourishment (*Silberberg and Silberberg*, 1940; *Lussier*, 1951; *Frandsen et al.*, 1954; *Bavetta et al.*, 1959; *Schneider and Adar*, 1964).

Figure 1 (*Dearden and Mosier*, 1974) presents photomicrographs of tibial epiphyseal cartilages, with the four growth zones indicated, from a rat fasted for 48 h, and from a well-fed control the same age.

The overall reduction in width of the epiphyseal cartilage is apparent. The decreased width of the epiphyseal cartilage is accompanied by an effective cessation of mitosis in the cartilaginous proliferative zone (*Saxton and Silberberg*, 1947; *Follis*, 1956; *Pratt and McCance*, 1960, 1964a), and chondrocyte atrophy (*Diatchenko*, 1897; *Silberberg and Silberberg*, 1940; *Frandsen et al.*, 1954;

Acheson, 1959). The width of the proliferative zone and the number of cells comprising it are reduced with PCM; this is similarly true for the zone of resting cartilage and particularly the zones of maturing cartilage and calcifying cartilage (*Silberberg and Silberberg*, 1940; *Saxton and Silberberg*, 1947; *Platt and Stewart*, 1962; *Pratt and McCance*, 1964a; *Deo et al.*, 1965; *Dearden and Mosier*, 1974).

In PCM, the zone of calcifying cartilage is characterized by reduced vascular invasion, with decreased osteoblastic activity and fewer osteoblasts (*Silberberg and Silberberg*, 1940; *Saxton and Silberberg*, 1947; *Frandsen et al.*, 1954; *Pratt and McCance*, 1964a; *Jha and Ramalingaswami*, 1968). The impaired vascular invasion results from a 'sealing off' of the epiphyseal cartilage by excess calcification at the metaphyseal margin (*Frandsen et al.*, 1954; *Bavetta and Bernick*, 1956; *Pratt and McCance*, 1964a). This excess calcification at the metaphyseal margin of the epiphyseal cartilage may be related to the transverse trabeculae or plaques reported in PCM (*Platt and Stewart*, 1962; *Platt et al.*, 1963; *Pratt and McCance*, 1964b; *Restrepo et al.*, 1964), and to the appearance of transverse striae, or lines of increased density. These lines are discussed in detail in a later section of the present paper. With PCM, newly formed trabeculae are few in number and irregularly shaped, having small cartilaginous cores, many appearing short, fragmented, or club-like (*Diatchenko*, 1897; *Handler et al.*, 1947; *Frandsen et al.*, 1954; *Pratt and McCance*, 1960, 1964a).

Appositional bone growth at periosteal and endosteal surfaces is altered in PCM. Mitosis in the undifferentiated cells is impaired, and there is considerable reduction in the number of osteoblasts (*Silberberg and Silberberg*, 1940; *Pratt and McCance*, 1960, 1964b; *Jha*, 1973), while the number of osteoclasts at the endosteal surfaces increases (*Pratt and McCance*, 1960; *El-Maraghi et al.*, 1965). On the surfaces of the trabeculae, the number of osteoblasts and osteoclasts are reduced, indicating little bone remodelling (*Silberberg and Silberberg*, 1940; *Pratt and McCance*, 1960, 1964b; *Jha* 1973). The compact bone of undernourished individuals is characterized by irregularly arranged fibers, abnormal Haversian remodeling, and enlarged vascular spaces surrounded by sclerotic cement lines (*Silberberg and Silberberg*, 1940; *Pratt and McCance*, 1960, 1964b; *Dickerson and McCance*, 1961; *Jha*, 1973). Also, alterations in bone histochemistry, and in the intracellular structure of bone cells are associated with PCM (*Likins et al.*, 1957; *Dearden and Mosier*, 1971, 1974; *Dearden and Espinosa*, 1974).

There is no histological evidence to indicate qualitative differences in the responses of bone to different protein-calorie deficiencies; for example, protein insufficiency in contrast to caloric deficiency, or starvation. Similarly, there is no indication that specific essential amino acid deficiencies produce any histological changes in bone that are qualitatively different from those described above for general protein deficiency. Histological changes very similar to those found with general protein deficiency have been reported in rats with diets deficient in lysine (*Harris et al.*, 1943; *Bavetta and Bernick*, 1955; *Haggar et al.*, 1955), phenylalanine (*Maun et al.*, 1945a; *Schwartz et al.*, 1951), tryptophan (*Scott*, 1955; *Bavetta and Bernick*, 1956), histidine (*Maun et al.*, 1946; *Scott*, 1954), threonine (*Scott and Schwartz*, 1953), and lucine (*Maun et al.*, 1954b); and diets containing chemical antagonists to phenylalanine (*Kaufman et al.*, 1961) and methionine (*Klavins et al.*, 1959). Growth of osteoblasts *in vitro* is suppressed in media deficient in those amino acids listed above, as well as glutamic acid, aspartic acid, proline, and cystine (*Fischer*, 1948). The most severe inhibition in growth of osteoblasts was observed with lysine deficiency.

III. Bone Composition

As with many experimental variables, results concerning chemical constituents of bone and bones and their interrelationships vary somewhat according to sample, methodology, etc. There are, however, some general patterns that are relatively consistent in most species, including man. Some changes in bone composition accompanying dietary deficiencies in protein *and* calories, and starvation are summarized in table I.

Generally speaking, in the course of normal post-natal bone development in most species, the concentrations of mineral and organic elements increase, while the relative contribution of water decreases (*Hammett*, 1925; *Dickerson*, 1962a, b; *Widdowson and Dickerson*, 1964). Thus, one may speak of chemical maturity of bone. With undernutrition due to deficiencies in protein *and* calories, bones generally tend to be less mature chemically than those of well-fed controls the same age (table I), although the differences are always small. The largest consistent alterations in constituent concentrations in bone with dietary deficiencies in protein *and* calories are a relative decrease in fat content, and a concomitant increase in percentage of water, on wet weight and fat-free weight bases. This was demonstrated early for cats by *Wellman* (1908), and for dogs by *Aron* (1911).

There is no deviation from normal in the Ca/P ratio in bone after underfeeding, although the Ca/N and Ca/collagen ratios are consistently greater in the cortical bone of these individuals. These latter patterns are not seen when whole bone is analyzed. The increased Ca/N and Ca/collagen ratios suggest the collagen is highly calcified. This is consistent with the findings of reduced organic matrix in bones of underfed rats when compared to normal controls (*Outhouse and Mendel*, 1933; *Zucker and Zucker*, 1946). The ratio (A:R) of ash weight of bone to that of the dry fat-free bone minus the ash (a measure of matrix), is similar to the Ca/collagen ratio, although the findings concerning the A:R ratio in pigs and

Table I. General changes in composition of bone with dietary deficiency in protein *and* calories, or starvation relative to well-fed controls of the same age

	Whole bone	Cortex	Epiphyses	References
Concentrations				
Calcium	slight decrease	slight decrease	slight decrease	1–3, 6, 9–11, 14
Phosphorus	no change	slight decrease	slight decrease	1, 2, 6, 9–11, 14
Total N	slight decrease	slight decrease	slight decrease	1, 2, 5, 9
Collagen N	slight decrease	slight decrease	slight decrease	1, 2
Ash	slight decrease	–	–	4, 7, 10, 14
Water	increase	–	–	1, 3, 5, 8, 12
Fat	decrease	–	–	1, 5, 12
Ratios				
Ca/N	slight decrease	increase	slight decrease	1–3
Ca/P	no change	no change	–	1, 9
Ca/collagen	no change	increase	slight decrease	1, 2, 13

Concentration: usually g/100 g dry, fat-free solids, or percentage of dry, fat-free weight. Water: g/100 g wet weight, or g/100 g fat-free weight. Fat: g/100 g dry weight; percentage of wet weight.
[1] *Dickerson and John* (1969). [2] *Dickerson and McCance* (1961). [3] *Dickerson and Widdowson* (1960). [4] *El-Maraghi et al.* (1965). [5] *Halliday* (1967). [6] *McCance* (1966). [7] *Platt and Stewart* (1966). [8] *Restrepo et al.* (1964). [9] *Shenolikar* (1966). [10] *Walker and Arvidsson* (1954). [11] *Weiske* (1897). [12] *Wellman* (1908). [13] *Widdowson* (1968, 1974). [14] *Yeager and Winters* (1935).

rats with PCM are equivocal (*Dickerson and McCance,* 1961; *Platt and Stewart,* 1962; *El-Maraghi et al.,* 1965).

It should be emphasized that the bone responses to undernutrition given in table I are only the most frequently reported, and there are conflicting findings. *Adams* (1971) compared bone composition of runt piglets with that of normal-sized littermates and found significantly higher mean concentrations of calcium and phosphorus, and significantly reduced mean Ca/collagen and Ca/N ratios in the humeral epiphysis of the runts, a pattern contrary to that usually reported (table I). It was assumed the runt piglets were small due to malnutrition *in utero.* As the data in table I are concerning changes in bone composition occurring with postnatal PCM, the alterations in bone composition seen in the runts may indicate some prenatal bone responses to PCM different from those occurring after birth, or suggest that runts are small due to prenatal factors other than typical PCM. *Restrepo et al.* (1964) analyzed femora and tibiae of young Guatemalan children who had died of PCM (marasmus and kwashiorkor). The reported fat concentrations were up to twice that expected for well-nourished children the same age (*Dickerson,* 1962b).

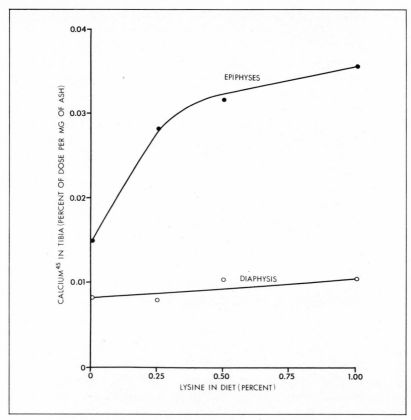

Fig. 2. Mean concentrations of ^{45}Ca in tibial diaphysis and epiphyses of rats on different levels of dietary lysine (*Likins et al.,* 1957).

There are few reports regarding changes in bone composition with protein deficiency, although some conclusions can be drawn. Organic matrix of bone may be reduced in protein deficiency in rats and monkeys (*Doyle and Porter,* 1951; *Jha et al.,* 1968), although *Harkness et al.* (1958) reported no change in the amount of collagen in femora of mice after 20 days on a protein-free diet. With protein deficiency in rats, there is little alteration in the concentrations of calcium, phosphorus or the Ca/P ratio in samples of cortex and epiphysis (*Yeager and Winters,* 1935; *Likins et al.,* 1957). Whole bone analyses indicate slightly less, or no difference in mean calcium concentration in bones from protein-deprived rats when compared to those from adequately nourished animals of the same age (*Yeager and Winters,* 1935; *Doyle and Porter,* 1951; *Shenolikar and Rao,* 1968). Accordingly, the mean percentage of ash may be less in the bones of protein-deficient rats (*Yeager and Winters*, 1935; *Doyle and Porter*, 1951; *El-Maraghi et al.*, 1965; *Shneolikar and Rao*, 1968), and pigs (*Platt and Stewart*, 1962). *Toverud and Toverud* (1933) found the mean percentage of calcium in the parietal bones and ribs of newly born babies was slightly greater for infants of mothers receiving a 'good' diet, including adequate milk during pregnancy, than for infants of mothers with deficient diets during pregnancy.

Likins et al. (1957) gave ^{45}Ca to rats being maintained on five levels of lysine. The mean concentrations of radiocalcium in the tibiae from the five groups of rats are shown in figure 2. It is apparent that an increase in dietary lysine was associated with increased deposition of radiocalcium. The effect was more pronounced in the epiphyses, particularly at low concentrations of lysine, than in the diaphyses; however, calcification is more active at the epiphyses, so the contrast between the two parts of the tibia is not surprising. If the effects of lysine on calcification are similar to those of general protein deficiency, as the histological evidence indicates, then protein adequacy seems an important factor in this part of normal chemical maturation of bone.

IV. The Weight of Bones and Skeleton

The attained weight of individual bones and the skeleton as a whole is less in starved or clinically malnourished children than in well-nourished children of the same age. *Dickerson and John* (1969) found the mean whole bone weights and fat-free weights of femora from 15 severely malnourished children (kwashiorkor and marasmus) significantly less than those of a normal control group of comparable age who had died of illness without apparent skeletal complications. The younger malnourished children (4—7.5 months) had relatively less bone weight than the older malnourished children (9—13 months). In four children diagnosed as 'atrophic' studied by *Ohlmüller* (1882), body weights were about 40% below that of normal children of comparable age and height; the mean fresh wet weight of bones was 14% below that of the normal children, and the mean dry fat-free skeletal weight was 9.3% less than the well-nourished children. These findings are consistent with the observations of *Kerpel-Fronius and Frank* (1949) who reported fresh whole skeletal weight of a malnourished child to be 15% less than expected for his age. Similarly, the total weight of bone mineral of malnourished children is less than that of well-nourished controls (*Gopalan et al.*, 1953; *Garrow and Fletcher*, 1964; *Garrow et al.*, 1965; *Halliday*, 1967).

Observations comparable to those for malnourished children have been made on several different experimental animals under various degrees of PCM: cats (*Voit*, 1866; *Sedlmair*, 1899; *Wellman*, 1908), cockerels (*Dickerson and McCance*, 1961), dogs (*Aron*, 1911), lambs (*Wallace*, 1948), mice (*Harkness et al.*, 1958), pigeons (*Chossat*, 1842), pigs (*McMeekan*, 1940; *Pomeroy*, 1941;

Dickerson and McCance, 1961; *McCance et al.,* 1962), and rats (*Jackson,* 1915; *Stewart,* 1916, 1918, 1919; *Winters et al.,* 1927; *McCay et al.,* 1935, 1939).

The depressed growth in bone weight seen in PCM can begin with the fetus *in utero* and may be associated with the level of maternal nutrition (*Barry,* 1920; *Wallace,* 1948), as well as apparent individual differences in the availability of nutrients to the fetus; an example of the latter is runt piglets of large litters (*Adams,* 1971; *Widdowson,* 1971), Weights of selected bones and groups of bones from 144 day-old twin lamb fetuses of ewes on 'high' and 'low' planes of nutrition during pregnancy were investigated by *Wallace* (1948). Growth in skeletal weight was clearly depressed in fetuses of underfed ewes for each bone and each group of bones; the degree of growth depression was fairly uniform between regional groups of bones, amounting to approximately 35% for those of the head, 37% for those of the neck, 41% for those of the thorax, 40% for those of the pelvis and shoulders, and 39% for those of the legs.

Although growth in skeletal weight is severely depressed in PCM, the bones continue to become heavier, though at a much slower rate than normal. This is true in experimental animals even if body weight is maintained at a constant level by underfeeding (*Aron,* 1911; *Jackson,* 1915, 1921; *Stewart,* 1916, 1918, 1919; *Jackson and Stewart,* 1918; *Dickerson and McCance,* 1961). Consequently, with malnutrition in animals (*Aron,* 1911; *Jackson,* 1915; *Stewart,* 1919), and man (*Jackson,* 1925), the skeleton becomes relatively heavier in proportion to body weight than in well-nourished individuals. When compared to normal individuals, the relative deficiency in skeletal weight in malnourished children is far less than that of total body weight or weight of the viscera (*Ohlmüller,* 1882; *Marfan,* 1921; *Kerpel-Fronius and Frank,* 1949; *Jackson,* 1925).

Growth in weight of individual bones and of the skeleton is depressed by experimental starvation (*Voit,* 1866; *Sedlmair,* 1899), protein deficiency (*Winters et al.,* 1927; *Limson and Jackson,* 1932; *Yeager and Winters,* 1935), caloric deficiency (*McCay et al.,* 1935; *Yeager and Winters,* 1935), deficiency in protein and calories (*Aron,* 1911; *McCay et al.,* 1939; *McMeekan,* 1940; *Dickerson and Widdowson,* 1960), and by protein deficiency with additional carbohydrate (*DiOrio et al.,* 1973). Comparing effects on bone weight of protein deficiency with those of caloric deficiency, starvation, or protein-calorie imbalance is extremely difficult between different studies because of the great variability in the timing, severity, duration, and exact nature of the dietary deficiencies; and the different animals, bones, and methods selected in various studies of experimental malnutrition. Similarly, this is true in studies of bone weight in human malnutrition, for the reasons outlined above, and because of variable terminology and uncertain etiologies of the various malnutritional syndromes.

Surprisingly few data are available for reliable comparisons of different protein-calorie deficiencies associated with changes in bone weight. *Winters et al.* (1927) maintained three groups of 25 young male rats at constant body weight for a period of 40 days following a period of 25–30 days of well-nourished growth. For the weight maintenance, three different dietary regimens were used: low calorie, low protein, and lysine-deficient. The dry weights of limb bones from these test animals were compared with those from well-nourished animals of the same body weight that were, of course, much younger. Compared to the younger, well-nourished animals, all groups of the malnourished animals had greater mean weights of the bones examined; i.e. the bones of the test animals continued to grow in weight beyond that attained before the weight mainte-nance diets. Unfortunately, no age controls were included in the study to allow comparisons with normal growth rates during the test period. These results showed little difference between quantitative protein deficiency (low protein) and qualitative protein deficiency (low lysine) in the absolute mean bone weights, or the increases in mean bone weight relative to well-fed rats of the same body weight. These data, however, indicate that in the calorie-deficient rats, mean growth in bone weight was depressed to a greater extent than for either of the protein-deficient groups. That dietary deficiencies in calories, or protein *and* calories, may depress growth in bone weight more than quantitative or qualitative protein deficiencies is consistent with the findings from several other studies (*Jackson*, 1925; *Yeager and Winters*, 1935; *DiOrio et al.*, 1973); however, there are also observations to the contrary (*Limson and Jackson*, 1932; *Pickens et al.*, 1940; *Armstrong*, 1948; *Cabak et al.*, 1963). There is some evi-dence in rats to suggest little or no protein, with extra carbohydrate, depresses growth in bone weight more than a low-protein diet (*DiOrio et al.*, 1973).

Interrelationships between protein and calories are extremely complex, partly because all protein can be used for calories; and with low-protein diets anorexia usually occurs, decreasing intake of non-protein calories. Many investi-gators are not sufficiently explicit in describing experimental methodology regarding PCM and bone to allow detailed comparisons between studies; how-ever, in general, the following seems to be true in experimental animals. When dietary restriction is below energy requirements for body maintenance, calories appear most important for bone growth, irrespective of the source. After energy levels above that required for body maintenance are met, protein appears most important for bone growth. Consequently, diets can be designed for experi-mental groups that allow either 'low calorie' or 'low protein' regimens to be most efficacious for bone growth when groups are compared, depending on the nature and degree of the deficiencies.

V. Bone Density

Many different measures of density have been applied to bone; however, most are designed to evaluate osteopenia. Frequently, osteopenia, the loss or absence of bone substance, has been diagnosed by increased radiolucency. Purely visual assessment of bone rarefaction from radiographs, however, has been shown to be unreliable, as at least 25–30% of the calcium normally present in bone tissue must be lost to be visually detectable on radiographs (*Lachman and Whelan*, 1936; *Ardran*, 1951); accordingly, radiological findings occur only in severe cases. Purely visual assessment of osteopenia cannot distinguish between bone absence due to osteoporosis and that due to narrow cortices. Nevertheless, visual assessment of bone rarefaction is common and is encountered frequently in reports concerning PCM.

Adams and Berridge (1969) subjectively graded the amount of trabecular bone in metacarpals from radiographs of 10 Ugandan children with kwashiorkor and 17 clinically healthy children; the trabecular bone was graded good, fair, or poor. Using these categories, there was significantly less trabecular bone in metacarpals of children with kwashiorkor than those of normal children. Visual examination of bone by *Spies et al.* (1953) in a group of 82 chronically under-nourished children showed that the amount of spongiosa in the metacarpals was below normal in 62% of the cases and the bone frequently had a cinder-like appearance. Unusually radiolucent areas were noted in bones of 68% of the children, and the 'quality' of compact bone was below average in the entire sample. Other investigators have reported increased radiolucency, abnormal trabeculation, and a cinder-like or ground-glass appearance of bones from radiographs of children with PCM (*Dean*, 1960; *El Nawaby et al.*, 1962; *Clement et al.*, 1964; *Garn et al.*, 1966a; *Ramos Galvan et al.*, 1969; *Maniar et al.*, 1974). *Reichman and Stein* (1968), however, found only 10% of a sample of children with severe PCM to have any rarefaction of bone, as assessed from radiographs of the chest, arm, and hand.

Reports of 'hunger osteoporosis', 'hunger osteomalacia', and 'hunger osteopathy' were frequent during the first half of this century; these came particularly from countries ravaged by the world wars. These osteopathies were variously described and are difficult to evaluate, but many appear to have resulted primarily from vitamin deficiencies rather than PCM *per se*, because the conditions often responded to vitamin therapy, particularly vitamin D, and not to realimentation (*Hume and Nirenstein*, 1921; *Burger et al.*, 1945; *Stare*, 1945). A detailed discussion of these hunger osteopathies may be found in *Keys et al.* (1950). Bone density is reduced with inflammatory bowel disease or following gastric surgery (*Morgan et al.*, 1965a; *Aukee et al.*, 1975; *Genant et al.*, 1976a, b); however, an etiology other than simple PCM must be considered, as the condition often responds to therapeutic doses of vitamin D (*Morgan et al.*, 1965a, b).

Radiographic densitometry or photodensitometry of bone expresses the amount of bone mineral in terms of an equivalent density: ivory, aluminum, etc. Radiographic densitometry is a valid technique for assessment of local bone mineral *in vivo*, and has been shown to be highly correlated (0.92–0.99) with bone ash and bone mineral (*Mack and Vogt*, 1968).

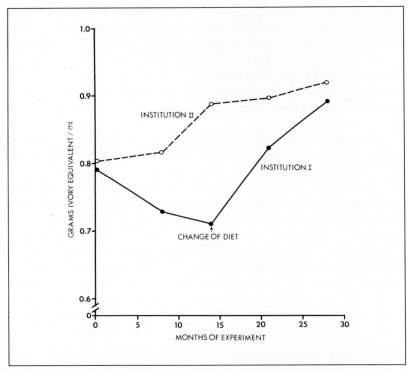

Fig. 3. Mean density of calcaneus of children in two institutions during dietary protein experiment (*Mack et al.*, 1947 c; *Mack*, 1949 a). See text for explanation.

Reddy et al. (1972) found the mean photodensity of femora of well-nourished rats to be greater than that of femora from rats maintained on low-protein diets of low-calorie diets. The mean bone density of the calorie-deficient rats was significantly lower than that of the protein-deficient rats.

Mack and her associates (*Mack et al.*, 1947 a, b, c; *Mack*, 1949 a, b) investigated changes occurring in bone density, and other biochemical and growth variables, of undernourished orphanage children while on two different dietary supplementation regimens. The supplemented diets of all children met recommended levels for sex and age and were matched as closely as possible for grams of protein and total energy. Children in one orphanage (institution I) received meat twice weekly, and the balance of the protein allowance was provided by legumes and other vegetable products; while children in institution II received meat ten times weekly. The net dietary difference between the two groups, then, was one of protein quality, not quantity. The two groups of children were studied for a period of 14 months, at which time the dietaries of the two institutions were reversed, and the children were studied for the subsequent 14 months. Part of their results concerning mean photodensity of the calcaneus are given in figure 3.

In the first experimental period, the higher meat group (institution II) showed significant increases in mean bone density, while the low meat group (institution I) had a progressive decline in mean density (fig. 3). With the diets reversed, children in institution I (now with the higher meat intake) showed marked increases in bone density, while the other children (institution II) showed only slight increases in mean density. These increases in bone density were not accompanied by any corresponding advances in skeletal maturity.

It seems clear from the foregoing that radiographic photodensity of bone is subnormal with protein deficiency, and that protein supplementation of deficient individuals is accompanied by increases in bone density. This has been reconfirmed in a subsequent study by *Mack et al.* (1962) and may also hold true for general PCM (*Schraer and Newman*, 1958; *El-Maraghi and Stewart*, 1963; *El-Maraghi et al.*, 1965). Some studies have found no associations between protein intake and radiographic photodensity; however, in these cases the basal diets of the subjects were apparently more than adequate (*Mainland*, 1963; *Odland et al.*, 1972).

The evidence concerning PCM and the density of bone determined gravimetrically is unclear. *McCay et al.* (1935, 1939) found reduced mean density of whole rat femora of calorie-restricted animals when compared to those of well-nourished controls, but increased specific gravity of whole rat humeri and femora with protein-calorie deficiency has been reported by *Saville and Lieber* (1969). *Jha et al.* (1968) reported greater density of femoral cortex of protein-deficient monkeys than that of well-fed animals. Some of these conflicting findings may arise from methodological differences, although it is difficult to understand increased gravimetric density of bone with PCM in light of overwhelming evidence to the contrary from other measures of bone density.

VI. Bone Growth: Elongation and Apposition

The effects of PCM on elongation are most clearly seen in the growth of tubular bones. Growth in length of tubular bones is depressed with prolonged fasting or starvation (*Carlson and Hoelzel*, 1947; *Quimby*, 1951), chronic deficiencies in protein *and* calories (*Outhouse and Mendel*, 1933; *Dickerson and McCance*, 1961), caloric deficiency (*McCay et al.*, 1935, 1939; *Adams*, 1969; *Kerr et al.*, 1973; *Fleagle et al.*, 1975), and protein deficiency (*Frandsen et al.*, 1954; *Heard et al.*, 1958; *Bunyard*, 1972; *Fleagle et al.*, 1975).

This depressed growth in length of tubular bones compared to well-nourished controls of the same age has been reported for children with severe

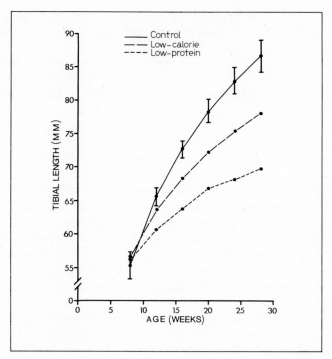

Fig. 4. Mean tibial length of three groups of *Cebus* monkeys maintained on low-calorie diet, low-protein diet, and well-nourished controls (± 1 SD) (*Fleagle et al.*, 1975).

and mild to moderate PCM (*Variot*, 1907; *Grande and Rof*, 1944; *Jones and Dean*, 1956; *Adams and Berridge*, 1969), and for a number of experimental animals: cockerels (*Pratt and McCance*, 1960, 1961), dogs (*Platt and Stewart*, 1968), monkeys (*Kerr et al.*, 1973; *Fleagle et al.*, 1975), pigs (*McMeekan*, 1940; *Dickerson and McCance*, 1961; *McCance et al.*, 1961; *Adams*, 1969), and rats (*Quinn et al.*, 1929; *Outhouse and Mendel*, 1933; *Saxton and Silberberg*, 1947; *Frandsen et al.*, 1954).

In experimental animals, newborn offspring of mothers restricted in protein, or protein *and* calories, have shorter mean limb bone lengths than do offspring of comparable gestational age from well-nourished mothers (*Platt and Stewart*, 1968; *Allen and Zeman*, 1971; *Dickerson and Hughes*, 1972; *Shrader and Zeman*, 1973; *Riopelle et al.*, 1976); the same is true of full-term runt piglets compared to average-sized littermates (*Adams*, 1971; *Widdowson*, 1971).

The extent of growth depression of bone length associated with PCM can be seen in figure 4, drawn from the serial data of *Fleagle et al.* (1975) for tibial length of *Cebus* monkeys. Groups of *Cebus* monkeys on three dietary regimens were studied serially for 196 days. One group (n = 12) was maintained on an

Table II. Mean rates of growth estimated for bone dimensions from individually fit curves from 8 to 28 weeks of life of monkeys on three dietary regimens[1,2]

	Dietary groups					
	control (n = 12)		low-protein (n = 14)		low-calorie (n = 6)	
	mean	SD	mean	SD	mean	SD
Humerus, length, mm	19.3	2.5	8.2	1.3	13.7	0.9
Radius, length, mm	18.4	2.4	8.3	1.4	12.4	1.4
Femur, length, mm	24.8	2.5	13.2	1.4	17.4	2.8
Tibia, length, mm	25.1	2.8	10.9	1.5	17.2	1.1
Femur, diameter, mm	1.1	0.2	0.3	0.1	0.6	0.9

[1] From *Fleagle et al.* (1975).
[2] Rates estimated as mean regression coefficients (b) from individual curves of the form $y = a + b (\log_e$ age in days)

adequate diet *ad libitum,* while another group (n = 4) was kept on a low-protein diet sufficient to maintain body weight at a relatively constant level; a third group (n = 6) was given a low-calorie diet (67% of control diet). After the age of 15 weeks, the differences from the means of the control animals are statistically significant ($p < 0.05$), as are the differences between mean tibial lengths for protein-deficient and those for calorie-deficient animals.

There are few data that allow estimations of rates of elongation with PCM. *Fleagle et al.* (1975), in the same experiment just described, fit individual curves to serial measurements of long bones from radiographs taken throughout the experimental period; the mean estimated rates of bone growth for the three groups are given in table II. The mean rate of growth in length of the major long bones is clearly depressed in the protein-deficient and calorie-deficient animals when compared to adequately nourished control animals, as is mean rate of growth in femoral diameter.

For all of the bone dimensions given in table II (*Fleagle et al.,* 1975), the mean rates of growth in the protein-restricted monkeys are considerably less than those of the animals growing under caloric restriction. In contrast, rats kept on low-calorie diets by *Winters et al.* (1927) showed generally greater gains in long bone lengths (on a percentage basis) than did protein-restricted animals, when compared to well-fed animals of the same body weight, but much younger. It is apparent that comparisons between the effects of protein deficiencies and caloric deficiencies are difficult to interpret; as expressed earlier, this is probably due to variation in the nature and degree of the deficiencies and perhaps differences among experimental animals. Examination of other studies comparing the effects of protein deficiency with those of caloric deficiency on bone elongation in experimental animals lead to similar equivocal conclusions (*Heard et al.,* 1958; *Adams,* 1969; *Kerr et al.,* 1973; *Stewart,* 1975).

Although growth in length is depressed with PCM, elongation continues at a reduced rate even when body weight of experimental animals is maintained at a constant level for an extended period through dietary restriction (*Winters et al.,* 1927; *Barnes et al.,* 1947; *Dickerson and Hughes,* 1968; *Dickerson and McCance,* 1961; *Pratt and McCance,* 1960, 1961, 1964b).

If malnourished animals are rehabilitated dietarily, bones respond with a catch-up period of rapid compensatory growth, although the more mature the bone at the time of rehabilitation, the smaller the catch-up, and the less the finally attained size (*Barnes et al.,* 1947). Similarly, if the period of malnutrition occurs relatively late in the growth period, the bone will be little affected by the nutritional insult, as adult size will already have been effectively attained. The latter is apparently why *Acheson and Macintyre* (1958) failed to find any significant effect of starvation on the elongation of relatively mature rat metacarpals and metatarsals.

The present evidence indicates when animals that are severely malnourished while young are rehabilitated, catch-up growth in bone length and the extended growing period are insufficient to allow attainment of normal adult bone lengths (*Barnes et al.,* 1947; *Pratt and McCance,* 1961; *McCance,* 1964, 1966; *Lister and McCance,* 1967; *Platt and Stewart,* 1968; *McCance and Widdowson,* 1962). What constitutes severe PCM in this regard is difficult to define, but the degree and duration of the malnutrition are important, as well as timing relative to the growth period. Rats malnourished *in utero* may show full recovery to normal values for some, but not all, long bone lengths if rehabilitated shortly after birth (*Dickerson and Hughes,* 1972).

As noted above (table II), the data of *Fleagle et al.* (1975) indicate slower mean rate of growth in femoral diaphyseal diameter of malnourished monkeys than that of adequately nourished animals. Using tetracycline labeling, *Shrader and Zeman* (1973) found rat pups from mothers deprived of protein during pregnancy to have a significantly slower mean rate of tibial periosteal growth than pups of well-nourished mothers. Impaired rates of periosteal growth with protein deficiency postnally can be seen in figure 5 (*Jha et al.,* 1968), which compares periosteal apposition during 6 weeks, visualized with tetracycline, for well-fed and protein-deficient monkeys. These findings are consistent with many studies documenting depressed growth in diameter of various tubular bones of malnourished children and experimental animals compared to well-nourished controls the same age (*McCay et al.,* 1935; *McFie and Welbourn,* 1962; *Widdowson and McCance,* 1963; *Adams,* 1969; *Allen and Zeman,* 1971; *Himes et al.,* 1975). Caution should be exercised when comparing bone widths between

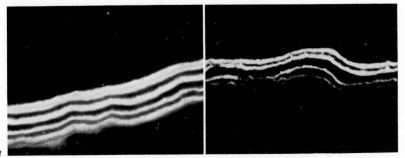

a

b

Fig. 5. Periosteal bone formation in a control monkey *(a)* and a protein-deficient monkey *(b)* visualized with tetracycline. Intervals between top two lines indicate 1 week's bone growth on stock diet. Intervals between subsequent lines reflect bone growth every 15 days on experimental diets. Ultraviolet light. × 160. Reprinted with permission from *Jha et al.* (1968).

populations because of genetic factors involved in these measures (*Smith et al.*, 1973).

Garn et al. (1969) and *Adams and Berridge* (1969) did not find significantly smaller metacarpal diaphyseal diameters in children hospitalized with kwashiorkor compared to local controls who themselves suffered from mild to moderate PCM. This suggests chronic protein-calorie deficiency is more important in depressing growth periosteally than the more acute, often rapidly precipitated bout with kwashiorkor. Children who are chronically malnourished due to celiac disease have significantly smaller metacarpal widths than healthy children of the same age (*Barr et al.*, 1972).

With protein deficiency in rats *in utero*, endosteal appositional growth is depressed significantly (*Shrader and Zeman*, 1973); this occurs also when monkeys are malnourished postnatally (*Jha et al.*, 1968). *Garn et al.* (1969) reported increased metacarpal medullary diameters in boys hospitalized with kwashiorkor compared to mild to moderately malnourished controls. When the boys with kwashiorkor were studied serially during recovery, the authors documented bone resorption at the endosteal surface and increases in metacarpal length (*Garn et al.*, 1966a). Some endosteal resorption during PCM is consistent with the histological evidence discussed previously concerning increased numbers of osteoclasts and fewer osteoblasts at the endosteal surfaces of tubular bones.

Depressed periosteal and endosteal apposition, and/or increased endosteal resorption in PCM results in thinner cortices of tubular bones. The amount of cortical bone is reduced in children and animals with PCM compared to that for well-nourished controls of the same age, whether the cortices are assessed visually, are measured directly from bones or radiographs, or the amount of cortex is estimated as volumes, cross-sectional areas, or by various indices (*Saxton and*

Silberberg, 1947; *Graham and Morales*, 1963; *Behar et al.*, 1964; *Garn et al.*, 1964a, b; *Restrepo et al.*, 1964; *Salomon et al.*, 1970; *Blanco et al.*, 1972a; *Shrader and Zeman*, 1972). Cortical thickness is also reduced in children chronically malnourished due to celiac disease (*McCrae and Sweet*, 1967; *Prader et al.*, 1969; *Barr et al.*, 1972).

It might be argued that smaller transverse bone dimensions, including cortical thickness, seen in malnourished children, reflect only smaller body size rather than a bone-specific response to nutritional stress. Transverse bone dimensions are, in fact, correlated with body size in well-nourished children and in children with mild to moderate PCM (*Reynolds*, 1944; *Maresh*, 1961, 1966; *Mazess and Cameron*, 1971; *Himes et al.*, 1976; *Yarbrough et al.*, 1977). Based on metacarpal dimensions, however, moderately malnourished children still have smaller transverse bone dimensions than would be expected for their stature and weight (*Himes et al.*, 1975).

Although growth in length, and growth in diameter of tubular bones are both inhibited by PCM, the proportions of the bones, as determined by width–length ratios, generally remain unaltered (*Marfan*, 1921; *Dickerson and Widdowson*, 1960; *Dickerson and McCance*, 1961; *Widdowson and McCance*, 1963; *Tonge and McCance*, 1965). *Shrader and Zeman* (1973), however, reported tibiae from young rats born of protein-restricted mothers to be disproportionately thin for their length compared to those of rats from adequately nourished mothers. The weight of limb bones relative to their length is less in malnourished animals than in well-nourished controls as measured by bone weight–length ratios (*McMeekan*, 1940; *Dickerson and McCance*, 1961) or robusticity indices (*Riesenfeld*, 1973). Less bone weight for length in PCM is consistent with evidence regarding increased endosteal resorption, failure to gain bone endosteally, and decreased density of cortical and trabecular bone.

For many bones of the skeleton, growth is much more complicated than that in tubular bones. In areas of the body such as the pelvis and craniofacial complex, a single bone dimension may reflect endochondral and appositional growth, and considerable bone remodeling and relocation. The mean length and breadth of the skull are less in malnourished animals than in well-nourished controls (*Dickerson and Widdowson*, 1960; *McCance et al.*, 1961; *Widdowson and McCance*, 1963; *El-Maraghi et al.*, 1965; *Dickerson and Hughes*, 1972). The flat bones of the skull in children with frank PCM are reduced in thickness, especially at bregma, vertex, lambda, and over the entire occiput between inion and foramen magnum (*Garn et al.*, 1966b). Similar thinning of the cranial vault has been observed in undernourished pigs (*McCance et al.*, 1961).

Craniofacial dimensions are not stunted uniformly with experimental malnutrition. *Tonge and McCance* (1965) maintained pigs on severe caloric restriction for 52 weeks. Table III gives the means of selected craniofacial dimensions of their undernourished pigs and the normal controls, and the means of the

Table III. Mean dimensions of undernourished pigs and of normal pigs the same age[1]

Dimension[2]	Undernourished (n = 5–9)	Normal (n = 2–4)	Percentage of normal
Body weight, kg	5.4	168	3.2
Upper facial length (basion-prosthion)	14.7	27.9	52.6
Lower facial length (basion-gnathion)	11.5	21.6	53.2
Facial height (nasal tip-interdentale superius)	2.3	4.8	47.9
Palate length (orale-staphylion)	8.9	16.9	52.6
Palate breadth (inner borders of sockets of the two upper second molars, endomolaria)	2.5	4.0	62.5
Mandible, condylar breadth (right-left condylion laterale)	7.7	14.7	52.4
Mandible, angular breadth (right-left gonia)	6.4	13.4	47.8
Mandible, anterior breadth (inner borders of the right-left mental foramina)	2.1	3.2	65.6
Mandible, length	12.7	25.2	50.4
Mandible, height of body (interdentale inferius-gnathion)	3.5	7.7	45.5
Mandible, height of condyle (gonion-highest points of condyle)	5.4	13.1	41.2
Width of ramus (least width at right angles to height)	4.6	6.9	66.7
Humerus, length	9.0	20.9	43.1
Humerus, breadth	1.6	3.7	43.2

[1] From *Tonge and McCance* (1965).
[2] All bone dimensions in centimeters.

undernourished animals expressed as percentages of those of the normals. For comparison, the same values are given for body weight, and length and breadth of humerus. For the craniofacial dimensions, the percentage of growth depression ranges from 41.2%, for mandibular height, to 66.7%, for width of ramus. The least affected craniofacial dimensions tended to be the breadths. Generally, craniofacial dimensions were affected somewhat less than those of the humerus. These findings are generally consistent with those for other pigs who were similarly undernourished (*McCance et al.,* 1961), and for underfed rats (*Outhouse and Mendel,* 1933; *Riesenfeld,* 1973), and puppies (*Platt and Stewart,* 1968). The small mandibular length in severely malnourished animals is accompanied by crowding and malocclusion of teeth; tooth size is less affected by PCM (*McCance et al.,* 1961; *Platt and Stewart,* 1962; *Tonge and McCance,* 1965, 1973; *Owens,* 1968; *Bunyard,* 1972).

Often human malocclusion has been attributed to poor diet (*Howe,* 1924; *Marshall,* 1932); however, the evidence comes primarily from animal experiments of severe PCM, such as those discussed above. *Parker et al.* (1952) investigated 14 maxillofacial dimensions and angles taken from cephalometric radiographs of 88 chronically malnourished children. The only consistent difference between the malnourished children and the normal reference data for children of the same race was slight subnormal inclination of the maxillary incisors, indicated by a low mean U1-NS angle. Examination of a large series of children with kwashiorkor showed little evidence of unusual malocclusion (*Trowell et al.,* 1954).

Mean dimensions of the pelvis are smaller in protein-calorie malnourished rats than in adequately nourished controls (*Outhouse and Mendel,* 1933; *Dickerson and Hughes,* 1968). *Dickerson and Hughes* (1972) found that newborn rat pups of mothers restricted in protein and calories during pregnancy had significantly smaller mean pelvic length, iliac length, and bi-iliac width than pups from well-nourished dams. Fetal malnutrition, however, did not affect mean bi-acetabular width. These same animals had significantly smaller mean cranial length and width, spinal length, and limb lengths than the controls. After 119 days of dietary rehabilitation, the mean pelvic and spinal dimensions were still significantly smaller than those of the controls who had not been malnourished, while the mean skull and limb dimensions were not significantly different from controls. Similar results to these were found for undernourished weanling rats who were later rehabilitated (*Dickerson et al.,* 1972). The authors suggest that the difference in response to dietary rehabilitation was due to differences among bones in their timing of the peak velocity of growth. The assumptions underlying this hypothesis are that bones are unable to fully catch up if rehabilitation occurs after the time of peak velocity growth, and/or the period of peak velocity growth is most conducive for full catch-up growth. Neither of these assumptions has been tested experimentally. The hypothesis regarding timing of peak velocity growth of bones and ability for catch-up, of course, cannot explain why bone dimensions were differentially affected by PCM initially.

While linear proportions of single bones generally remain unaltered in PCM, size relationships between bones do change due to differential effects of malnutrition. *Riesenfeld* (1973) maintained rats on diets severely restricted in protein and calories for 6 months and found the lengths of pelvis, and of the long limb bones, were significantly reduced relative to the length of the neurocranium; while relative to spinal length, lengths of pelvis and long limb bones were larger than those of normal rats. That spinal length is relatively more affected by PCM than long bone lengths in animals is supported by evidence from other investigators (*Platt and Stewart*, 1968; *Dickerson and Hughes*, 1972; *Shrader and Zeman*, 1973).

VII. Skeletal Maturation

Skeletal maturity is based on various 'maturity indicators' relating to onset of ossification, and subsequently, primarily shape changes in the developing skeleton. Usually, skeletal maturity is determined from radiographs of various areas of the body, most usually the hand-wrist or knee, although skeletal maturation can also be determined histologically, chemically, etc.

The timing of onset of ossification, as determined by the mean or median ages of appearance of radiopaque ossification centers, is delayed in children with mild to moderate PCM (*Dreizen et al.*, 1958, 1959, 1961; *Chávez et al.*, 1964; *Ghosh et al.*,, 1966, 1967; *Frisancho et al.*, 1970; *Blanco et al.*, 1972b; *Salomón et al.*, 1972; *Maniar et al.*, 1974), and in children with kwashiorkor or marasmus (*El Nawaby et al.*, 1962; *Massé*, 1962; *Massé and Hunt*, 1963; *Clement et al.*, 1964; *Maniar et al.*, 1974), when compared to reference data for adequately nourished children of the same sex. *Shrader and Zeman* (1972, 1973) found significant delay in onset of ossification in rat pups of malnourished dams, while *Dickerson and Widdowson* (1960) found significant delay in onset of ossification in young rats undernourished postnatally.

In a study of chronically malnourished Alabama children, *Dreizen et al.* (1958) found the hand-wrist centers most frequently delayed relative to reference times of appearance (*Greulich and Pyle*, 1950) were trapezium and metacarpals III and IV (94%) for boys; and trapezium (90%) and distal ulna (89%) for girls. The least frequently delayed centers in this series were adductor sesamoid (47%) and distal radius (59%) for boys; and proximal phalanges III (54%), IV and V (59%) for girls.

Trying to assess which hand-wrist ossification centers were most sensitive to malnutrition, *Frisancho et al.* (1970) calculated percentage delay in median age of onset of each center and groups of centers, corrected for gestational age, for moderately malnourished children in Guatemala relative to median ages of well-nourished US children (*Garn et al.*, 1967). On a percentage basis, the most delayed hand-wrist centers were triquetral for boys (63%) and girls (71%); followed by distal radius and proximal phalanx V (49%) for boys, and lunate (64%) and metacarpal I (52%) for girls. A similar analysis by *Blanco et al.* (1972b) for a different sample of malnourished Guatemalan children showed the hand-wrist centers with the greatest percentage delay were in boys triquetral (60%) and lunate (52%); and lunate (49%) and triquetral (43%) in girls.

A better approach to the assessment of differential sensitivity of ossification centers to PCM is expression of delay in time of onset of ossification in terms of standard scores. These scores take into account the normal variability in the ages of onset of ossification and express deviations from the reference means in standard deviation units. *Salomón et al.* (1972) determined the mean standard score deviations of times of appearance of hand-wrist centers of moderately malnourished Guatemalan children relative to those of *Stuart et al.* (1962) for US children. In this series, the most delayed centers were metacarpal II, middle phalanx III, and proximal phalanx IV for boys; and proximal phalanx III and metacarpal II for girls.

Todd (1931, 1937) stated variability in the pattern and rate at which ossification centers appear is due to ill health, poor nutrition, or unhygienic conditions. This position was supported by a number of workers (*Francis, 1939, 1940; Francis and Werle, 1939; Pyle et al., 1948; Greulich and Pyle, 1959*), although their conclusions were drawn primarily from clinical observations. Accordingly, the sequence of appearance of ossification centers has been described are regular and constant for healthy children. Differential effects of malnutrition on the times of appearance would then lead to alterations in the order of appearance of centers of malnourished children (*Greulich, 1954*). Although such notions may seem reasonable, there is considerable variation in sequence of appearance of ossification centers of the hand and wrist of well-nourished healthy children (*Pryor, 1907; Baer and Durkatz, 1957; Garn and Rohmann, 1960a; Garn et al., 1961, 1966a, 1972*); further, the variation present has a significant genetic component (*Pryor, 1907; Reynolds, 1943; Garn et al., 1963*). Consequently, attributing variation in order of onset of ossification to protein-calorie malnutrition should be done with caution, and must take into account the variability in adequately nourished individuals.

Unfortunately, sequences of appearance of ossification centers in malnourished children have primarily been described as sequences of mean times of appearance from cross-sectional samples, rather than mean or median sequences of appearance, which can be determined only from longitudinal data. The sequence of mean times hides any variability in the order of ossification and precludes identification and assessment of frequencies of unusual sequences.

Chávez et al. (1964) report the largest single deviation from the *Greulich and Pyle* (1959) sequence of hand-wrist centers for malnourished Mexican children was that of the triquetral; the same was true of malnourished boys from Alabama (*Dreizen et al., 1964a*) and Guatemala (*Garn et al., 1966c*). However, the marked variability of the triquetral, even in the absence of malnutrition, is well known (*Johnston et al., 1968; Poznanski et al., 1971*). Available evidence for order of mean times of appearance of ossification centers of the hand and wrist in moderately malnourished children show some differences in ossification order within groups of bones, e.g. carpals and metacarpals, compared to those established for well-nourished children (*Dreizen et al., 1964a; Ghosh et al., 1966, 1967*).

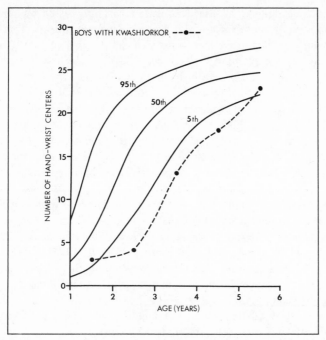

Fig. 6. Median number of hand-wrist ossification centers of Guatemalan boys with kwashiorkor (*Garn et al.,* 1966) compared to percentiles for well-nourished US boys (*Garn and Rohmann,* 1960b).

Garn et al. (1966c) compared the sums of differences between ranks of order of median times of appearance for 28 hand-wrist centers from moderately malnourished Guatemalan children, and the median sequence of well-nourished Ohio children (*Garn and Rohmann,* 1960a). The summed differences between ranks were 32 for boys and 28 for girls. Comparable sums for differences between ranks of different samples of well-nourished children from the US range from 4 to 18 for boys and 4 to 26 for girls (*Garn and Rohmann,* 1960a). By similar comparison of the median sequence of appearance of hand-wrist centers of moderately malnourished Guatemalan boys (*Yarbrough et al.,* 1973) with that for US boys (*Garn and Rohmann,* 1960a), one obtains a sum of differences between ranks of 29. When the sequence (median?) of chronically malnourished Alabama children (*Dreizen et al.,* 1964a) is compared to those for well-nourished Ohio children (*Garn and Rohmann,* 1960a), the sum values of differences between ranks are 32 for boys and 14 for girls. These comparisons suggest that there may be some small total variation in median sequence in the hand and wrist due to malnutrition *per se* in boys, but none in girls, beyond what might be expected due to genetic or sampling differences.

The rate of appearance of ossification centers can be assessed by simply counting the number of centers present at a particular point in time. This has been most frequently applied to the hand-wrist area and is useful only until most of the centers have appeared; about 7 years of age in boys and 6 years in girls. Although there is considerable variation in number of hand-wrist centers at a given age in well-nourished children (*Garn and Rohmann,* 1960b), the median number of hand-wrist centers of malnourished children are significantly smaller than those of well-nourished children the same age. This can be seen in figure 6, which presents median numbers of hand-wrist centers of Guatemalan boys with kwashiorkor (*Garn et al.,* 1966b) relative to percentiles for well-nourished US boys (*Garn and Rohmann,* 1960b). The boys with kwashiorkor approximate or fall below the fifth percentile of US boys from 1 through $5\frac{1}{2}$ years of age.

Smaller mean numbers of hand-wrist centers compared to well-nourished children have been reported for several groups of children with mild to moderate or severe PCM (*Franics,* 1940; *Béhar et al.,* 1964; *Rohmann et al.,* 1964; *Guzman et al.,* 1965; *Blanco et al.,* 1972b; *Yarbrough et al.,* 1973). *Sontag and Wines* (1947) found no significant associations between number of ossification centers at several sites of newborns and infants, and protein intake of the mothers during pregnancy. All of the mothers, however, were well nourished, and it is unlikely that any deficiency effects could be detected at the levels of protein intake described (lowest category, up to 55 g/day).

Significantly fewer centers of ossification are present in newborn rat pups of dams restricted in protein during pregnancy than in pups of comparable gestational age from well-nourished dams (*Shrader and Zeman,* 1973).

Béhar et al. (1964) and later *Garn et al.* (1966b) report comparisons of number of hand-wrist centers of Guatemalan children with kwashiorkor, with local children with mild to moderate PCM. The number of hand-wrist centers of the children with kwashiorkor was not significantly different from the local controls, although local controls were below the fifth percentile for number of centers for US children; further, second metacarpal cortical thickness was significantly less for the children with kwashiorkor than for local controls. This suggests that the rate of appearance of hand-wrist centers is affected more by chronic protein-calorie deficiency than the acute malnutritional state, while cortical thickness appears sensitive to rather short-term nutritional stress.

Garn et al. (1966b) did not find any evidence of catch-up growth in the number of hand-wrist centers in severely malnourished children during the first year of nutritional rehabilitation. More prolonged studies of nutritional supplementation, however, show moderately malnourished children may significantly increase the mean number of hand-wrist centers relative to control children at the end of a 5-year program (*Guzman et al.,* 1964, 1965, 1968).

Skeletal maturation, expressed as bone age or skeletal age relative to reference standards for well-nourished children, is retarded in children with mild to

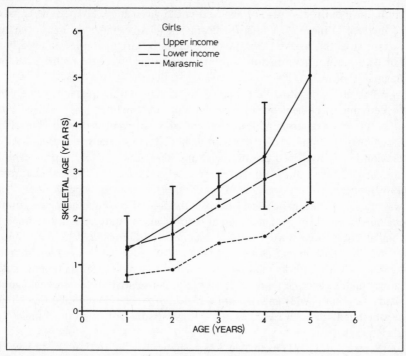

Fig. 7. Mean skeletal ages of girls with marasmus compared to those of girls from low-income and high-income (± 1 SD) strata in India (*Maniar et al.,* 1974).

moderate PCM (*Zayaz et al.,* 1940; *Spies et al.,* 1953, 1959; *Dreizen et al.,* 1954, 1961, 1967; *Acheson et al.,* 1962; *Salomón et al.,* 1972). Similarly, clinically malnourished children, usually with marasmus and/or kwashiorkor, are retarded in skeletal age relative to reference standards, and relative to local controls, who themselves are usually mildly to moderately malnourished (*Gopalan,* 1956; *Jones and Dean,* 1956, 1959; *Massé,* 1962; *Graham and Morales,* 1963; *Massé and Hunt,* 1963; *Graham,* 1967, 1968; *Stini,* 1969; *Satgé et al.,* 1970). Skeletal maturation is retarded also in children malnourished due to celiac disease (*Prader et al.,* 1963, 1969; *Shmerling et al.,* 1968; *Rey et al.,* 1971; *Barr et al.,* 1972), inflammatory bowel disease, and regional enteritis (*Sobel et al.,* 1962; *McCaffery et al.,* 1970).

Figure 7 presents mean Greulich-Pyle skeletal ages from cross-sectional samples of Indian girls from two income levels and girls with nutritional marasmus (*Maniar et al.,* 1974). Girls with marasmus were significantly retarded in skeletal age relative to both income groups of girls; these latter girls were, themselves, retarded relative to the standards at most ages. The differences between the mean skeletal ages of the marasmic girls and those of the standard range from 2.6 months at 1 year of age to over 2.5 years at age 5. The general pattern of increase in mean skeletal age in marasmic girls is similar to that of the lower income girls; this suggests that the marasmic girls started retarded at an early age, failed to catch up, but matured skeletally at a similar rate as girls from the lower income stratum.

Several investigators have studied children who were treated for severe PCM to determine whether complete catch-up in bone maturity has occurred. *Briers et al.* (1975) examined African children who had recovered from clinical PCM 8–14 years previously. Compared to local children of the same age and sex, the previously malnourished children were not significantly different in mean bone age delay (*Tanner et al.,* 1962) or metacarpal cortical index (*Barnett and Nordin,* 1960), although they were significantly shorter than the local controls. There was no significant association between the age of previous hospitalization for malnutrition and current metacarpal cortical index, stature, or lag in bone age. Similar conclusions were drawn by *Suckling and Campbell* (1957) and *Krueger* (1969). *MacWilliam and Dean* (1965), however, found little catch-up in skeletal age of children with kwashiorkor during the first year after treatment; indeed, some children became more skeletally retarded relative to the skeletal age standards. *Graham* (1968) and *Graham and Adrianzen* (1971) reported that children rehabilitated from kwashiorkor for 2–8 years did not catch up completely in skeletal age.

Keet et al. (1971) examined 123 ex-patients who previously had kwashiorkor, 5 and 10 years after recovery, and compared them with unaffected siblings. The bone age (*Greulich and Pyle,* 1959) of the ex-patients was more variable at a given chronological age than that of their sibs, and estimation of median skeletal ages from the graphic presentations suggests some persistent skeletal retardation after 5- and 10-year periods. Unfortunately, no values for mean or median skeletal ages were presented in the paper.

Studies of skeletal maturation from radiographs in animals equivalent to skeletal age in man are few because of lack of reliable standards. *Dickerson and Hughes* (1972) maintained female rats on protein-calorie-deficient diets throughout pregnancy. Newborn pups from the undernourished dams had significantly lower mean skeletal maturity scores (*Hughes and Tanner,* 1970) than pups from well-nourished dams; however, after 140 days of nutritional rehabilitation, the young rats malnourished *in utero* did not differ in mean bone maturity from the controls. *Riopelle et al.* (1976), however, report little consistent or significant effect on skeletal maturity of newborn rhesus monkeys of different levels of maternal protein intake during pregnancy. Postnatal PCM retards skeletal maturation in rats (*Jackson,* 1925; *Silberberg and Silberberg,* 1957; *Acheson and Macintyre,* 1958; *Dickerson et al.,* 1972), and monkeys (*Thurm et al.,* 1976); and, at least in rats, catch-up in bone maturity can be effected with dietary rehabilitation (*Acheson and Macintyre,* 1958; *Dickerson et al.,* 1972).

There are few data concerning the effects of PCM on the epiphyseal fusion of bones. *Dreizen et al.* (1957) and *Dreizen and Stone* (1962) examined the ages at arbitrary stages of fusion of 21 bones of the hand and wrist for a sample of Alabama youths who were chronically malnourished. These individuals were compared to clinically normal children from the same area. For the malnourished individuals, epiphyseal fusion was delayed an average of slightly more than 1 year, compared to the control group. The delay was quantitatively similar in boys and girls and was not related to stature.

Growth in size of epiphyses and round bones is not skeletal maturation *per se,* but is correlated with maturational status and is often used as an index of bone maturation. *Tompkins and Wiehl* (1954) did not find significant differences in the mean dimensions of the distal femoral epiphyses and proximal tibial epiphyses of infants from mothers receiving a protein supplement during pregnancy compared to infants from unsupplemented mothers. The unsupplemented mothers, however, were adequately nourished and had an average protein intake of 76 g/day during the last half of pregnancy. *Frandsen et al.* (1954) found the anteroposterior depth of the proximal tibial epiphysis positively associated with the protein intake of young rats, although no such association was demonstrated for the distal tibial epiphysis. The mean transverse diameter of the distal femoral epiphysis of presumably malnourished runt piglets is significantly smaller than that of normal-sized littermates (*Adams,* 1971), as are the mean dimensions of proximal tibial epiphyses of moderately malnourished Spanish children compared to US children of the same age (*Grande and Rof,* 1944).

Several investigators· have found the mean projected total areas of the carpal bones, as determined from radiographs (*Carter,* 1926), less in moderately malnourished children than in well-nourished children of the same age and sex (*Zayaz et al.,* 1940; *MacNair and Roberts,* 1938; *Roberts and MacNair,* 1937; *Berridge and Prior,* 1954). This ossification index exhibits a compensatory catch-up in malnourished children after a period of milk supplementation (*MacNair and Roberts,* 1938; *Roberts and MacNair,* 1937).

VIII. Lines of Increased Density and Ossification Anomalies

Lines of increased density have variously been called lines of arrested growth, stress lines, bone scars, transverse lines or striae, and Harris lines. Whether these lines result from completely arrested bone growth, slowed growth, or from growth recovery is the basis of some controversy (*Harris,* 1931; *Park and Richter,* 1953; *Acheson,* 1959; *Park,* 1964). The lines are found in other than tubular bones (*Garn et al.,* 1968b; *Caffey,* 1972), and in these instances they are seldom transverse. Regardless of nomenclature, these lines, which are parallel to the growing surfaces of bones and distinguished by in-

Table IV. Percentage prevalence of lines of increased density in bones of malnourished children[1]

Reference	Area examined	Kwashiorkor		Marasmus		Mild to moderate malnutrition	
		boys	girls	boys	girls	boys	girls
Jones and Dean (1956)	hand-wrist	66[2] (53)	–	–	–	25[2] (41)	–
Jones and Dean (1959)	knee	84 (44)	74 (31)	–	–	22 (23)	10 (24)
Restrepo et al. (1964)	knee	50[2] (18)	–	20[2] (5)	–	0[2] (5)	–
Maniar et al. (1974)[3]	hand-wrist	14 (19)	21 (26)	30 (79)	28 (66)	22 (435)	22 (385)

[1] Numbers in parentheses indicate number in sample.
[2] Sexes combined.
[3] Estimated from graph.

creased mineralization, may be distinguished histologically or radiographically. The histology of these lines has been discussed by a number of authors (*Eliot et al.*, 1927; *Harris*, 1933; *Follis and Park*, 1952; *Park and Richter*, 1953; *Acheson*, 1959, 1960; *Park*, 1954, 1964).

There is reasonably good evidence that frequency of lines of increased density and incidence of new lines in bones are associated with childhood illness (*Hewitt et al.*, 1955; *Acheson and Macintyre*, 1958; *Acheson*, 1959; *Marshall*, 1968; *Garn et al.*, 1968b). Illness is apparently sufficient to effect lines of increased density, but it is not, however, necessary.

Lines of increased density have been produced in bones by starvation in rabbits (*Asada*, 1924), puppies (*Harris*, 1933), and rats (*Acheson*, 1959). *Park and Richter* (1953) observed lines of increased density in the growing ends of femora, tibiae, and ribs of young rats fed protein-free diets. *Stewart and Platt* (1958) found lines of increased density in the distal metaphyses of radii of pigs on protein-deficient diets, and protein-deficient diets with added carbohydrate (*Platt and Stewart*, 1962).

Lines of increased density are reported to be common in association with severe (*Jones and Dean*, 1956; 1959; *Dean*, 1960; *Clement et al.*, 1964; *Adams and Berridge*, 1969), and mild to moderate (*Grande and Rof*, 1944; *Dreizen et al.*, 1956, 1961, 1964b; *Salomón et al.*, 1972; *Blanco et al.*, 1974) PCM in children. *Higginson* (1954), however, found no evidence of lines at the costochondral junctions of ribs of 20 Bantu children who died of kwashiorkor.

Cross-sectional studies of clinically malnourished children generally indicate they have lines of increased density more frequently than local controls, who are themselves generally inadequately nourished. The prevalence of lines in bones of malnourished children from several studies is presented in table IV. The high degree of variability in the frequencies of individuals exhibiting lines of increased density may be due to different nutritional status or the transient nature of lines; although complications from concomitant disease cannot be ruled out, given the sensitivity of lines to illness, and the frequent accompaniment of illness with PCM.

Dreizen et al. (1956, 1964b) compared frequency of lines of increased density and the mean number of lines in the distal 10 mm of the radius of moderately malnourished children from Alabama, and of apparently healthy children from the same area. There was no significant differences between the two groups in the frequency of individuals having lines, nor in the mean number of lines per individual. However, the well-nourished children had more incomplete lines than the malnourished children. The authors used this as suggestive of better bone remodeling in the well-nourished children.

Lines of increased density may be rather transient (*Eliot et al.*, 1927; *Harris*, 1931; *Garn et al.*, 1968b), due probably to bone remodeling; although some lines may be retained into the ninth decade (*Garn and Schwager*, 1967). Consequently, longitudinal study is necessary to evaluate properly nutritional involvement in the genesis of lines of increased density; unfortunately, there are no serial data for development of lines in children with PCM. *Garn et al.* (1968a), however, followed serially a group of well-nourished Ohio children and state 'nearly all' had one or more lines on the distal aspect of the tibia at some time in their life. Understanding more fully the extent of sensitivity and specificity of lines of increased density to PCM must await appropriate longitudinal study.

A variety of anomalous ossification centers have been reported present in malnourished children. *Snodgrasse et al.* (1955) compared the number of epiphyseal anomalies (supernumery epiphyses, pseudoepiphyses, and notches) and non-epiphyseal anomalies present in the metacarpals and phalanges of the hand from radiographs of 142 malnourished Alabama children with those of 124 apparently normal children. There was a statistically significant difference in the mean number of epiphyseal anomalies between the malnourished children and the well-nourished children, although the actual difference between the means was small (malnourished children's mean, 1.94; healthy children's mean, 1.23). From these cross-sectional data, the frequency of malnourished children with anomalous ossification centers was 79% for boys and girls, while the corresponding frequencies for well-nourished boys and girls were 72 and 49%, respectively. There was no difference between the two groups of children in the mean number of non-epiphyseal anomalies. *Garn et al.* (1966b) noted a higher frequency of multiple foci of ossification in the epiphysis of the first metacarpal in Guatemalan children hospitalized with kwashiorkor and marasmus than for local healthy children of the same age and sex.

That anomalous ossification centers result from PCM *per se* should be accepted with reserve, given findings in adequately nourished children. Multiple foci of ossification are extremely common in healthy, well-nourished children (*Roche and Sunderland,* 1959; *Caffey,* 1972) and have been reported also in association with several diseases and abnormal genotypes (*Brailsford,* 1943; *Caffey,* 1972). *Lee and Garn* (1967) found all of 228 children who were studied serially to have notching in the non-epiphyseal ends of metacarpals I, II, and V at some time in their lives. *Posener et al.* (1939) found 96% of healthy children examined between 4 and 8 years of age in London to have pseudoepiphyses of the metacarpal bones. It has been suggested that metacarpal and phalangeal notching, and associated anomalies in children with PCM are related to relatively retarded skeletal maturation (*Snodgrasse et al.,* 1955; *Dreizen et al.,* 1965). *Levine* (1972), however, found no significant association between notching in metacarpals and phalanges, and rate of skeletal maturation in white children from South Africa.

Some other skeletal anomalies seem more positively linked to PCM. *Warkany and Nelson* (1940, 1941) examined skeletons of newborn rats from dams reared on protein-deficient diets since shortly after birth. Compared to rat pups from adequately nourished dams, the frequency of rat pups with skeletal abnormalities was more than 100 times greater in rats from protein-malnourished dams. Other than expected growth depression in bone dimensions, common skeletal abnormalities were: absence of radius, ulna, tibia, and fibula; syndactylism of hands and feet; fused ribs; and many irregularly shaped bones. Abnormal twisting of tibiae has been reported in arginine deficiency in chicks (*Newberne et al.,* 1960).

IX. Conclusions and Implications for PCM in Man

The effects of PCM on bone growth and development have been described. It is clear that several aspects of bone growth and development are altered in PCM, and are sensitive to protein-calorie deficiency, prenatally and during postnatal growth. Although much emphasis is placed on the depressing, retarding effects of PCM on bone growth and development, bone growth continues, though slowed, even under the most severe dietary restrictions; ossification centers still appear, though delayed; and epiphyses still fuse, though at a relatively later time. Further, in PCM there are few major alterations in bone composition, and there is little evidence for alteration of sequence of onset of ossification, or occurrence of certain ossification anomalies beyond what might be expected due to sampling or population differences.

In the present discussion, possible physiological mechanisms concerning PCM and bone have not been mentioned. There is little sure knowledge in this regard, although hormones are almost certainly involved. Some new work is being done documenting endocrine and other metabolic changes in PCM. Hopefully, these investigations will shed some light on mechanisms of bone responses to PCM.

Many of the data concerning PCM and bone are from animal studies. These provide experimental control and techniques usually unavailable in human studies; thus, animal studies are critical to this body of science. Moreover, from the point of view of basic science, and animal husbandry and agricultural science, findings on experimental animals are ends in themselves. Nevertheless, the exact applicability to humans of results from experimental animals is unknown. In many cases, the correspondence between the findings from animal and human studies seems reasonably good, e.g. histological changes in bone of monkeys and man resulting from kwashiorkor-like syndromes. In other instances, the exact implications of animal studies to human beings are unclear. The effects of different rates of bone growth and maturation, lengths of growth period, modes of locomotion, etc., are difficult to evaluate, especially in light of varying nutritional requirements.

Many have attempted to use bone growth and development to aid in the identification of children with mild to moderate PCM. Although many aspects of bone growth and development are sensitive to PCM, there is little evidence to indicate that they are specific to PCM. Consequently, measures of bone growth and development, like other measures of growth and maturation, are only supportive in evaluating nutritional status, and should be used in concert with other indicators of nutritional status, namely biochemistry, clinical assessment, dietary intake.

The data are limited, but suggest bone density and metacarpal cortical thickness may be as sensitive as body weight to dietary realimentation of malnourished children, and more sensitive than stature or skeletal age. If this could be substantiated, bone density and cortical thickness could prove very useful in documenting progress with dietary treatment following PCM.

It is difficult to determine whether the responses of bone to PCM are beneficial adaptations for malnourished individuals. Clearly, slowed growth allows additional nutrients for body maintenance; however, unlike most other body tissues, bones continue to grow and mature even during severe dietary restriction. Whether the mechanical integrity maintained through this continued growth is more beneficial than conservation of nutrients expended in growth is difficult to assess. Resorption of bone from endosteal and trabecular surfaces during PCM provides a mineral pool for the body, and a negligible amount of protein from resorbed collagen; however, much of this pool is lost through excretion due to negative metabolic balance of nitrogen and minerals.

There are no data reported concerning the effects in adulthood of PCM in childhood, or its amelioration. These concerns include questions regarding adult stature, susceptibility to fractures, senile osteoporosis, etc. Most of these questions can be answered only by effective longitudinal studies. Some recent analytical approaches, such as prediction of adult stature, offer promising alternatives to life-time serial studies, and may also be used effectively to evaluate skeletal growth and maturational effects with dietary rehabilitation. Finally, there has been much duplication in animal and human studies. Clearly, some replication is essential, but much time, effort, and money has been spent imprudently by repeating experiments, the outcomes of which could have been predicted.

Acknowledgments

Preparation of the manuscript was supported by grants HD-04629 and HD-10246 from the National Institutes of Health. Thanks are extended to *L.C. Dearden* and *G.J. Jha* for their kind permission to reproduce figures 1 and 5, and to *Molly Schwinn* for preparation of the manuscript.

References

Acheson, R.M.: Effects of starvation, septicaemia and chronic illness on the growth cartilage plate and metaphysis of the immature rat. J. Anat. *93:* 123–130 (1959).

Acheson, R.M.: Effects of nutrition and disease on human growth; in *Tanner* Human growth. Symp. Soc. Hum. Biol., vol. 3, pp. 73–92 (Pergamon Press, Oxford 1960).

Acheson, R.M.: Maturation of the skeleton; in *Falkner* Human development, pp. 465–502 (Saunders, Philadelphia 1966).

Acheson, R.M.; Fowler, G.B., and Janes, M.D.: Effect of improved care on the predicted adult height of undernourished children. Nature, Lond. *194:* 735–736 (1962).

Acheson, R.M. and Macintyre, M.N.: The effects of acute infection and acute starvation on skeletal development. Br. J. exp. Path. *39:* 37–45 (1958).

Adams, P.: The effect of experimental malnutrition othe development of long bones. Biblthca Nutr. Dieta, No. 13, pp. 69–73 (Karger, Basel 1969).

Adams, P.H.: Intra-uterine growth retardation in the pig. Biol. Neonate *19:* 341–353 (1971).

Adams, P. and Berridge, F.R.: Effects of kwashiorkor on cortical and trabecular bone. Archs Dis. Childh. *44:* 705–709 (1969).

Allen, L.H. and Zeman, F.J.: Influence of increased postnatal food intake on body composition of progeny of protein-deficient rats. J. Nutr. *101:* 1311–1318 (1971).

Ardran, G.M.: Bone destruction not demonstrable by radiography. Br. J. Radiol. *24:* 107–109 (1951).

Armstrong, W.D.: Influence of nutritional factors on skeletal atrophy from disuse and on normal bones of mature rats. J. Nutr. *35:* 597–609 (1948).

Aron, H.: Nutrition and growth. I. Philip. J. Sci. *6:* 1–55 (1911).

Asada, T.: Über die Entstehung und pathologische Bedeutung der im Röntgenbild des Rohrenknochens am Diaphysenende zum Vorschein kommenden 'parallelen Querlinienbildung'. Mitt. Med. Fak. Univ. Kyushu Fukuoka *9:* 43–95 (1924).

Aukee, S.; Alhava, E.M., and Karjalainen, P.: Bone mineral after partial gastrectomy. II. Scand. J. Gastroent. *10:* 165–169 (1975).

Baer, M.J. and Durkatz, J.: Bilateral asymmetry in skeletal maturation of the hand and wrist: a roentgenographic analysis. Am. J. phys. Anthrop. *15:* 181–196 (1957).

Barnes, L.L.; Sperling, G., and McCay, C.M.: Bone growth in normal and retarded growth rats. J. Geront. *2:* 240–243 (1947).

Barnett, E. and Nordin, B.E.C.: The radiological diagnosis of osteoporosis: a new approach. Clin. Radiol. *11:* 166–174 (1960).

Barr, D.G.D.; Shmerling, D.H., and Prader, A.: Catch-up growth in malnutrition, studied in celiac disease after institution of gluten-free diet. Pediat. Res. *6:* 521–527 (1972).

Barry, L.W.: The effects of inanition in the pregnant albino rat, with special reference to the changes in the relative weights of the various parts, systems and organs of the offspring. Carnegie Inst. (Wash.) Contrib. Embryol. *53:* 91–136 (1920).

Bavetta, L.A. and Bernick, S.: Lysine deficiency and dental structures. J. Am. dent. Ass. *50:* 427–433 (1955).

Bavetta, L.A. and Bernick, S.: Effect of tryptophan deficiency on bones and teeth of rats. Oral Surg. *9:* 308–315 (1956).

Bavetta, L.A.; Bernick, S., and Ershoff, B.H.: Effects of caloric restriction on the bones and periodontium in rats. Archs Path. *68:* 630–638 (1959).

Béhar, M.: Prevalence of malnutrition among preschool children of developing countries; in *Scrimshaw and Gordon* Malnutrition, learning and behavior, pp. 30–41 (MIT Press, Cambridge 1968).

Béhar, M.; Rohmann, C.; Wilson, D.; Viteri, F., and Garn, S.M.: Osseous development in children with kwashiorkor. Fed. Proc. Fed. Am. Socs exp. Biol. *23:* 338 (1964).

Bengoa, J.M.: The problem of malnutrition. Wld Hlth Org. Chron. *28:* 3–7 (1974).

Berridge, F.R. and Prior, K.M.: The skeletal development of the children at the beginning and end of the period of experimental feeding. Med. Res. Counc. spec. Rep.. Ser. *287:* appendix D, pp. 119–130 (1954).

Blanco, R.A.; Acheson, R.M.; Canosa, C., and Salomón, J.B.: Sex differences in retardation of skeletal development in rural Guatemala. Pediatrics *50:* 912–915 (1972a).

Blanco, R.A.; Acheson, R.M.; Canosa, C., and Salomón, J.B.: Retardation in appearance of ossification centers in deprived Guatemalan children. Hum. Biol. *44:* 525–536 (1972b).

Blanco, R.A.; Acheson, R.M.; Canosa, C., and Salomón, J.B.: Height, weight, and lines of arrested growth in young Guatemalan children. Am. J. phys. Anthrop. *40:* 39–48 (1974).

Bourne, G.H.: The biochemistry and physiology of bone; 2nd ed. (Academic Press, New York 1971).

Brailsford, J.F.: Variations in the ossification of the bones of the hand. J. Anat. *77:* 170–175 (1943).

Briers, P.J.; Hoorweg, M., and Stanfield, J.P.: The long-term effects of protein energy malnutrition in early childhood on bone age, bone cortical thickness and height. Acta paediat. scand. *64:* 853–858 (1975).

Bunyard, M.W.: Effects of high sucrose cariogenic diets with varied protein-calorie levels on the bones and teeth of the rat. Calcif. Tiss. Res. *8:* 217–227 (1972).

Burger, G.C.E.; Sandstead, H.R., and Drummond, J.: Starvation in western Holland: 1945. Lancet *ii:* 282–283 (1945).

Cabak, V.; Dickerson, J.W.T., and Widdowson, E.M.: Response of young rats to deprivation of protein or calories. Br. J. Nutr. *17:* 601–616 (1963).

Caffey, J.: Pediatric X-ray diagnosis; 6th ed., vol. 2 (Year Book, Chicago 1972).

Carlson, A.J. and Hoelzel, F.: Growth and longevity of rats fed omnivorous and vegetarian diets. J. Nutr. *34:* 81–96 (1947).

Carter, T.M.: Technique and devices used in radiographic study of the wrist bones in children. J. educ. Psychol. *17:* 237–247 (1926).

Chávez, A.; Hidalgo, C.P. y Pitol, A.: Maduración ósea en dos grupos de niños con diferente estado de nutrición. Salud Publ., Mex. *6:* 705–717 (1964).

Chossat, M.: Note sur le système osseux. C. r. hebd. Séanc. Acad. Sci., Paris *14:* 451–454 (1842).

Clement, R.; Gounelle, H. et Reddy, K.-D.: Malnutrition et développement osseux de l'enfant. Presse méd. *72:* 225–227 (1964).

Dean, R.F.A.: The effects of malnutrition on the growth of young children. Mod. Probl. Paediat., vol. 111–112 (Karger, Basel 1960).

Dearden, L.C. and Espinosa, T.: Comparison of mineralization of the tibial epiphyseal plate in immature rats following treatment with cortisone, prophylthiouracil or after fasting. Calcif. Tiss. Res. *15:* 93–110 (1974).

Dearden, L.C. and Mosier, H.D.: Electron microscopy of tibial cartilage in catch-up growth. Anat. Rec. *169:* 304 (1971).

Dearden, L.C. and Mosier, H.D.: Growth retardation and subsequent recovery of the rat tibia, a histochemical, light, and electron microscopic study. II. After fasting. Growth *38:* 277–294 (1974).

Deo, M.G.; Sood, S.K., and Ramalingaswami, V.: Experimental protein deficiency. Archs Path. *80:* 14–23 (1965).

Diatchenko, E.: Sur les modifications dans la croissance des os des fœtus intra-utérins des lapines sous l'influence de l'inanition complète de leurs mères. C. r. 12e Congr. Méd., Moscow 1897, vol. 2, pp. 297–298.

Dickerson, J.W.T.: The effect of development on the composition of a long bone of the pig, rat and fowl. Biochem. J. *82:* 47–55 (1962a).

Dickerson, J.W.T.: Changes in the composition of the human femur during growth. Biochem. J. *82:* 56–61 (1962b).

Dickerson, J.W.T. and Hughes, P.C.R.: The effect of undernutrition on the growth and maturation of the skeleton of the rat. Proc. Nutr. Soc. *27:* 43a–44a (1968).

Dickerson, J.W.T. and Hughes, P.C.R.: Growth of the rat skeleton after severe nutritional intrauterine and post-natal retardation. Resuscitation *1:* 163–170 (1972).

Dickerson, J.W.T.; Hughes, P.C.R., and McAnulty, P.A.: The growth and development of rats given a low-protein diet. Br. J. Nutr. *27:* 527–536 (1972).

Dickerson, J.W.T. and John, P.M.V.: The effect of protein-calorie malnutrition on the composition of the human femur. Br. J. Nutr. *23:* 917–924 (1969).

Dickerson, J.W.T. and MacCance, R.A.: Severe undernutrition in growing and adult animals. 8. The dimensions and chemistry of the long bones. Br. J. Nutr. *15:* 567–576 (1961).

Dickerson, J.W.T. and Widdowson, E.M.: Some effects of accelerating growth. II. Skeletal development. Proc. R. Soc. B *152:* 207–217 (1960).

DiOrio, L.P.; Miller, S.A., and Navia, J.M.: The separate effects of protein and calorie malnutrition on the development and growth of rat bones and teeth. J. Nutr. *103:* 856–863 (1973).

Doyle, M.D. and Porter, T.E.: The effect of kind and level of protein in the diet of the production of soft and skeletal tissues. J. Nutr. *45:* 29–46 (1951).

Dreizen, S.; Currie, C.; Gilley, E.J., and Spies, T.D.: Observations on the association between nutritive failure, skeletal maturation rate and radiopaque transverse lines in the distal end of the radius in children. Am. J. Roentg. *76:* 482–487 (1956).

Dreizen, S.; Snodgrasse, R.M.; Dreizen, J.G., and Spies, T.D.: Seasonal distribution of initial appearance of postnatal ossification centers in hand and wrist of undernourished children. J. Pediat. *55:* 738–743 (1959).

Dreizen, S.; Snodgrasse, R.M.; Parker, G.S.; Currie, C., and Spies, T.D.: Maturation of bone centers in hand and wrist of children with chronic nutritive failure. Am. J. Dis. Child. *87:* 429–439 (1954).

Dreizen, S.; Snodgrasse, R.M.; Webb-Peploe, H., and Spies, T.D.: The effect of prolonged nutritive failure on epiphyseal fusion in the human hand skeleton. Am. J. Roentg. *78:* 461–470 (1957).

Dreizen, S.; Snodgrasse, R.M.; Webb-Peploe, H., and Spies, T.D.: The retarding effect of protracted undernutrition on the appearance of the postnatal ossification centers in the hand and wrist. Hum. Biol. *30:* 253–264 (1958).

Dreizen, S.; Spirakis, C.N., and Stone, R.E.: Chronic undernutrition and postnatal ossification. Am. J. Dis. Child. *108:* 44–52 (1964a).

Dreizen, S.; Spirakis, C.N., and Stone, R.E.: The influence of age and nutritional status on 'bone scar' formation in the distal end of the growing radius. Am. J. phys. Anthrop. *22:* 295–306 (1964b).

Dreizen, S.; Spirakis, C.N., and Stone, R.E.: The distribution and disposition of anomalous notches in the non-epiphyseal ends of human metacarpal shafts. Am. J. phys. Anthrop. *23:* 181–188 (1965).

Dreizen, S.; Spirakis, C.N., and Stone, R.E.: A comparison of skeletal growth and maturation in undernourished and well-nourished girls before and after menarche. J. Pediat. *70:* 256–263 (1967).

Dreizen, S. and Stone, R.E.: Human nutritive and growth failure. Postgrad. Med. *32:* 381–386 (1962).

Dreizen, S.; Stone, R.E., and Spies, T.D.: The influence of chronic undernutrition on bone growth in children. Postgrad. Med. *29:* 182–193 (1961).

Eliot, M.M.; Souther, S.P., ad Park, E.A.: Transverse lines in X-ray plates of the long bones of children. Bull. Johns Hopkins Hosp. *41:* 364–388 (1927).

El-Maraghi, N.R.H.; Platt, B.S., and Stewart, R.J.C.: The effect of the interaction of dietary protein and calcium on the growth and maintenance of the bones of the young, adult and aged rats. Br. J. Nutr. *19:* 491–508 (1965).

El-Maraghi, N.H.R. and Stewart, R.J.C.: The interaction of protein and calcium on the growth and composition of bones in young rats. Proc. Nutr. Soc. *22:* 30–31 (1963).

Fischer, A.: Amino acid metabolism of tissue cells *in vitro.* Nature, Lond. *161:* 1008 (1948).

Fleagle, J.G.; Samonds, K.W., and Hegsted, D.M.: Physical growth of cebus monkeys, *Cebus albifrons,* during protein or calorie deficiency. Am. J. clin. Nutr. *28:* 246–253 (1975).

Follis, R.B.: Some observations on experimental bone disease; in *Wolstenholme and O'Connor* Bone structure and metabolism, pp. 249–257 (Churchill, London 1956).

Follis, R.B. and Park, E.A.: Some observations on bone growth, with particular respect to zones and transverse lines of increased density in the metaphysis. Am. J. Roentg. *68:* 709–724 (1952).

Francis, C.C.: Factors influencing appearance of centers of ossification during early childhood. Am. J. Dis. Child. *57:* 817–830 (1939).

Francis, C.C.: Factors influencing appearance of centers of ossification during early childhood. Am. J. Dis. Child. *59:* 1006–1012 (1940).

Francis, C.C. and Werle, P.P.: The appearance of centers of ossification from birth to five years. Am. J. phys. Anthrop. *24:* 273–299 (1939).

Frandsen, A.M.; Nelson, M.M.; Sulon, E.; Becks, H., and Evans, H.M.: The effects of various levels of dietary protein on skeletal growth and endochondral ossification in young rats. Anat. Rec. *119:* 247–261 (1964).

Frisancho, A.R.; Garn, S.M., and Ascoli, W.: Unequal influence of low dietary intakes on skeletal maturation during childhood and adolescence. Am. J. clin. Nutr. *23:* 1220–1227 (1970).

Garn, S.M.; Béhar, M.; Rohmann, C.; Viteri, F., and Wilson, D.: Catch-up bone development during treatment of kwashiorkor. Fed. Proc. Fed. Am. Socs exp. Biol. *23:* 338 (1964a).

Garn, S.M.; Guzman, M.A., and Wagner, B.: Superiosteal gain and endosteal loss in protein-calorie malnutrition. Am. J. phys. Anthrop. *30:* 153–155 (1969).

Garn, S.M.; Hempy, H.O., and Schwager, P.M.: Measurement of localized bone growth employing natural markers. Am. J. phys. Anthrop. *28:* 105–108 (1968a).

Garn, S.M. and Rohmann, C.G.: Variability in the order of the bony centers of the hand and wrist. Am. J. phys. Anthrop. *18:* 219–230 (1960a).

Garn, S.M. and Rohmann, C.G.: The number of hand-wrist centers. Am. J. phys. Anthrop. *18:* 293–299 (1960b).

Garn, S.M.; Rohmann, C.G.; Béhar, M.; Viteri, F., and Guzman, M.A.: Compact bone deficiency in protein-calorie malnutrition. Science *145:* 1444–1445 (1964b).

Garn, S.M.; Rohmann, C.G., and Blumenthal, T.: Ossification sequence polymorphism and sexual dimorphism in skeletal development. Am. J. phys. Anthrop. *24:* 101–115 (1966a).

Garn, S.M.; Rohmann, C.G., and Davis, A.A.: Genetics of hand-wrist ossification. Am. J. phys. Anthrop. *21:* 33–40 (1963).

Garn, S.M.; Rohmann, C.G., and Guzman, M.A.: Malnutrition and skeletal development in the pre-school child; in Pre-school child malnutrition, pp. 43–62 (National Academy of Science National Research Council, Washington 1966b).

Garn, S.M.; Rohmann, C.G., and Silverman, F.N.: Radiographic standards for postnatal ossification and tooth calcification. Med. Radiogr. Photogr. *43:* 45–66 (1967).

Garn, S.M.; Rohmann, C.G., and Wallace, D.K.: Association between alternate sequences of hand-wrist ossification. Am. J. phys. Anthrop. *19:* 361–364 (1961).

Garn, S.M.; Sandusky, S.T.; Miller, R.L., and Nagy, J.M.: Developmental implications of dichotomous ossification sequences in the wrist region. Am. J. phys. Anthrop. *37:* 111–116 (1972).

Garn, S.M. and Schwager, P.M.: Age dynamics of persistent transverse lines in the tibia. Am. J. phys. Anthrop. *27:* 375–378 (1967).

Garn, S.M.; Silverman, F.N.; Hertzog, K.P., and Rohmann, C.G.: Lines and bands of increased density. Med. Radiogr. Photogr. *44:* 58–89 (1968b).

Garn, S.M.; Velonis, D.C.M.L.; Wiggins, P.; Lee, M.M.C.; Rohmann, C.G., and Eyman, M.: Ossification delay and sequence variability in Guatemala. Continuing Progress Report 66-2, part III, on PH-43-65-1006 (Fels Research Institute, Yellow Springs 1966c).

Garrow, J.S. and Fletcher, K.: The total weight of mineral in the human infant. Br. J. Nutr. *18:* 409–412 (1964).

Garrow, J.S.; Fletcher, K., and Halliday, D.: Body composition in severe infantile malnutrition. J. clin. Invest. *44:* 417–425 (1965).

Genant, H.K.; Mall, J.C.; Lanzl, L.H.; Horst, J. vander, and Wagonfeld, J.B.: Quantitative bone mineral analyses in patients with inflammatory bowel disease. Am. J. Roentg. *126:* 1303–1304 (1976a).

Genant, H.K.; Mall, J.C.; Wagonfeld, J.B.; Horst, J. vander, and Lanzl, L.H.: Skeletal demineralization and growth retardation in inflammatory bowel disease. Investve Radiol. *11:* 541–549 (1976b).

Ghosh, S.; Bhardawaj, O.P., and Varma, K.P.S.: A study of skeletal maturation of hand and wrist and its relationship to nutrition. Indian Pediat. *3:* 145–152 (1966).

Ghosh, S.; Varma, K.P.S., and Bhardawaj, O.P.: A study of skeletal maturation of hand and wrist and its relationship to nutrition. Indian Pediat. *4:* 11–20 (1967).

Gopalan, C.: Kwashiorkor in Uganda and Coonoor. J. trop. Pediat. *1:* 206–219 (1956).

Gopalan, C.; Venkatachlam, P.S., and Srikantia, S.G.: Body composition in nutritional edema. Metabolism *2:* 335–343 (1953).

Graham, G.G.: Effect of infantile malnutrition on growth. Fed. Proc. Fed. Am. Socs exp. Biol. *26:* 139–143 (1967).

Graham, G.G.: The later growth of malnourished infants, effects of age, severity, and subsequent diet; in *McCance and Widdowson* Calorie deficiencies and protein deficiencies, pp. 301–314 (Little, Brown, Boston 1968).

Graham, G.G. and Adrianzen, B.: Growth, inheritance, and environment. Pediat. Res. *5:* 691–697 (1971).

Graham, G.G. and Morales, E.: Studies in infantile malnutrition. I. Nature of the problem in Peru. J. Nutr. *79:* 479–487 (1963).

Grande, F. y Rof, J.R.: Alimentación y desarrollo infantil. III. Communicación Estudio radiografico del desarrollo esqueletico de un grupo de niños en edad escolar. Revta clin. esp. *12:* 234–240 (1944).

Greulich, W.W.: The relationship of skeletal status to the physical growth and development of children; in *Boell* Dynamics of growth processes, pp. 212–223 (Princeton University Press, Princeton 1954).

Greulich, W.W. and Pyle, S.I.: Radiographic atlas of skeletal development of the hand and wrist, p. 190 (Stanford University Press, Stanford 1950).

Greulich, W.W. and Pyle, S.I.: Radiographic atlas of skeletal development of the hand and wrist: 2nd ed., p. 256 (Stanford University Press, Stanford 1959).

Guzman, M.A.; Flores, M.; Bruch, H.; Salomón, J.B. y Béhar, M.: Efecto de la suplementación proteico-calorica en el desarrollo óseo de niños preescolares. Guatemala Pediat. *5:* 45–52 (1965).

Guzman, M.A.; Rohmann, C.; Flores, M.; Garn, S.M., and Scrimshaw, N.S.: Osseous growth of Guatemalan children fed a protein-calorie supplement. Fed. Proc. Fed. Am. Socs exp. Biol. *23:* 338 (1964).

Guzman, M.A.; Scrimshaw, N.S.; Bruch, H.A., and Gordon, J.E.: Nutrition and infection field study in Guatemalan villages, 1959–1964. Archs envir. Hlth *17:* 107–118 (1968).

Haggar, R.; Kinney, T.D., and Kaufman, N.: Bone healing in lysine-deficient rats. J. Nutr. *47:* 305–317 (1955).

Halliday, D.: Chemical composition of the whole body and individual tissues of two Jamaican children whose death resulted primarily from malnutrition. Clin. Sci. *33:* 365–370 (1967).

Ham, A.H.: Histology; 7th ed. (Lippincott, Philadelphia 1974).

Hammett, F.S.: A biochemical study of bone growth. I. Changes in the ash, organic matter, and water during growth *(Mus norvergicus albinus).* J. biol. Chem. *64:* 709–728 (1925).

Handler, P.; Baylin, G.J., and Follis, R.H.: The effects of caloric restriction on skeletal growth. J. Nutr. *34:* 677–689 (1947).

Harkness, M.L.R.; Harkness, R.D., and James, D.W.: The effect of a protein-free diet on the collagen content of mice. J. Physiol., Lond. *144:* 307–313 (1958).

Harris, H.A.: Lines of arrested growth in the long bones in childhood. The correlation of histological and radiographic appearances in clinical and experimental conditions. Br. J. Radiol. *4:* 561–588, 622–640 (1931).

Harris, H.A.: Bone growth in health and disease (Oxford University, London 1933).

Harris, H.A.; Neuberger, A., and Sanger, F.: Lysine deficiency in young rats. Biochem. J. *37:* 508–513 (1943).

Heard, C.R.C.; Platt, B.S., and Stewart, R.J.C.: The effects on pigs of a low-protein diet with and without additional carbohydrate. Proc. Nutr. Soc. *17:* xli-xlii (1958).

Hewitt, D.; Westropp, C.K., and Acheson, R.M.: Oxford child health survey. Effect of childish ailments on skeletal development. Br. J. prev. soc. Med. *9:* 179–186 (1955).

Higginson, J.: Studies on human bone from South African Bantu subjects. II. Histopathological changes in the ribs of South African Bantu infants. Metabolism *3:* 392–399 (1954).

Himes, J.H.: Nutrition problems of lesser-developed countries; in *Johnston* Anthropology, human nutrition and behavior (University of New Mexico Press, Albuquerque, in press).

Himes, J.H.; Malina, R.M., and Stepick, C.D.: Relationships between body size and second metacarpal dimensions in Oaxaca (Mexico) school children 6 to 14 years of age. Hum. Biol. *48:* 677–692 (1976).

Himes, J.H.; Martorell, R.; Habicht, J.-P.; Yarbrough, C.; Malina, R.M., and Klein, R.E.: Patterns of cortical bone growth in moderately malnourished preschool children. Hum. Biol. *47:* 337–350 (1975).

Howe, P.R.: Some experimental effects of deficient diets on monkeys. J. Am. dent. Ass. *11:* 1161–1165 (1924).

Hughes, P.C.R. and Tanner, J.M.: The assessment of skeletal maturity in the growing rat. J. Anat. *106:* 371–402 (1970).

Hume, E.M. and Nirenstein, E.: Comparative treatment of cases of hunger-osteomalacia in Vienna, 1920, as outpatients. Lancet *201:* 849–853 (1921).

Jackson, C.M.: Changes in the relative weights of the various parts, systems and organs of young albino rats held at constant body-weight by underfeeding for various periods. J. exp. Zool. *19:* 99–156 (1915).

Jackson, C.M.: The effects of various types of inanition upon growth and development, with special reference to the skeleton. Anat. Rec. *21:* 68–69 (1921).

Jackson, C.M.: The effects of inanition and malnutrition upon growth and structure (Blakiston, Philadelphia 1925).

Jackson, C.M. and Stewart, C.A.: The effects of underfeeding and refeeding upon the growth of the various systems and organs of the body. Minnesota Med. *1:* 403–414 (1918).

Jelliffe, D.B.: The assessment of the nutritional status of the community. Wld Hlth Org. Monogr. Ser. No. 53 (WHO, Genève 1966).

Jha, C.J.: Effect of protein depletion and repletion on bone turnover. Indian J. anim. Res. *7:* 4–8 (1973).

Jha, G.J. and Ramalingaswami, V.: Bone growth in protein deficiency. Am. J. Path. *53:* 1111–1123 (1968).

Johnston, F.E.; Whitehouse, K.H., and Hertzog, K.P.: Normal variability in the age and first onset of ossification of the triquetral. Am. J. phys. Anthrop. *28:* 97–99 (1968).

Jones, P.R.M. and Dean, R.F.A.: The effects of kwashiorkor on the development of the bones of the hand. J. trop. Pediat. *2:* 51–68 (1956).

Jones, P.R.M. and Dean, R.F.A.: The effects of kwashiorkor on the development of the bones of the knee. J. Pediat. *54:* 176–184 (1959).

Kaufman, N.; Klavins, J.V., and Kinney, T.D.: Pathologic changes induced by β-2-thienylalanine. J. Nutr. *75:* 93–103 (1961).

Keet, M.P.; Moodie, A.D.; Wittmann, W., and Hansen, J.D.L.: Kwashiorkor: a prospective ten-year follow-up study. S. Afr. med. J. *45:* 1427–1449 (1971).

Kerpel-Fronius, E. und Frank, K.: Einige Besonderheiten der Körperzusammensetzung und Wasserverteilung bei der Säuglingsatrophie. Ann. paediat., Basel *173:* 321–330 (1949).

Kerr, G.R.; Waisman, H.A.; Allen, J.A.; Wallace, J., and Scheffler, G.: Malnutrition studies in *Macaca mulatta.* II. The effect on organ size and skeletal growth. Am. J. clin. Nutr. *26:* 620–630 (1973).

Keys, A.; Brozek, J.; Henschel, A.; Nickelsen, O., and Taylor, H.L.: The biology of human starvation, vol. 1 (University of Minnesota Press, Minneapolis 1950).

Klavins, J.V.; Kinney, T.D., and Kaufman, N.: The effect of *DL*-ethionine on skeletal growth in rats. J. Nutr. *67:* 363–379 (1959).

Krueger, R.H.: Some long-term effects of severe malnutrition in early life. Lancet *ii:* 514–517 (1969).

Lachman, E. and Whelan, M.: The roentgen diagnosis of osteoporosis and its limitations. Radiology *26:* 165–177 (1936).

Lee, M.M.C. and Garn, S.M.: Pseudoepiphyses or notches in the non-epiphyseal end of metacarpal bones in healthy children. Anat. Rec. *159:* 263–272 (1967).

Levine, E.: Notches in the non-epiphyseal ends of the metacarpals and phalanges in children of four South African populations. Am. J. phys. Anthrop. *36:* 407–416 (1972).

Likins, R.C.; Bavetta, L.A., and Posner, A.S.: Calcification in lysine deficiency. Archs Biochem. Biophys. *70:* 401–412 (1957).

Limson, M. and Jackson, C.M.: Changes in the weights of various organs and systems of young rats maintained on a low-protein diet. J. Nutr. *5:* 163–174 (1932).

Lister, D. and McCance, R.A.: Severe undernutrition in growing and adult animals. 17. The ultimate results of rehabilitation: pigs. Br. J. Nutr. *21:* 787–799 (1967).

Lussier, J.-P.: Modifications histologiques du cartilage de conjugaison chez le rat au cours du jeûne protéique. Revue canad. Biol. *10:* 33–41 (1951).

Mack, P.B.: Comparison of meat and legumes in a controlled feeding program. IV. Second period with regimens reversed – dietaries and results of physical observations. J. Am. diet. Ass. *25:* 848–857 (1949a).

Mack, P.B.: Comparison of meat and legumes in a controlled feeding program. VI. Seven-month study of twenty-four children. J. Am. diet. Ass. *25:* 1017–1021 (1949b).

Mack, P.B.; Shevock, V.D., and Tomassetti, M.R.: Comparison of meat and legumes in a controlled feeding program. I. Dietary plan. J. Am. diet. Ass. *23:* 488–496 (1947a).

Mack, P.B.; Shevock, V.D., and Tomassetti, M.R.: Comparison of meat and legumes in a controlled feeding program. II. Medical, dental, and laboratory observations. J. Am. diet. Ass. *23:* 588–599 (1947b).

Mack, P.B.; Shevock, V.D., and Tomassetti, M.R.: Comparison of meat and legumes in a controlled feeding program. III. Discussion of findings. J. Am. diet. Ass. *23:* 677–685 (1947c).

Mack, P.B. and Vogt, F.B.: Assessment of bone mass by roentgenographic density techniques; in *Whedon and Cameron* Progress in methods of bone mineral measurement, pp. 27–80 (US Department of Health, Education and Welfare, Washington 1968).

Mack, P.B.; Vose, G.P.; Kinard, C.L., and Campbell, H.B.: Effects of lysine-supplemented diets on growth and skeletal density of preadolescent children. Am. J. clin. Nutr. *11:* 255–262 (1962).

MacNair, V. and Roberts, J.J.: Effect of a milk supplement on the physical status of institutional children. II. Ossification of the bones of the wrist. Am. J. Dis. Child. *56:* 494–509 (1938).

MacWilliam, K.M. and Dean, R.F.A.: The growth of malnourished children after hospital treatment. E. Afr. med. J. *42:* 297–303 (1965).

Mainland, D.: X-Ray bone density of infants in a prenatal nutrition study. Milbank Memorial Fund Q. *61:* 6–106 (1963).

Maniar, B.M.; Kapur, P.L., and Seevai, M.H.: Effect of malnutrition on bones of hand in children. Indian Pediat. *11:* 213–226 (1974).

Maresh, M.M.: Bone, muscle and fat measurements. Longitudinal measurements of the bone, muscle and fat widths from roentgenograms of the extremities during the first six years of life. Pediatrics *28:* 971–984 (1961).

Maresh, M.M.: Changes in tissue widths during growth. Roentgenographic measurements of bone, muscle and fat widths from infancy through adolescence. Am. J. Dis. Child. *111:* 142–155 (1966).

Marfan, A.B.: Les états de dénutrition dans la première enfance. Description de l'hypothrepsie et l'arthrepsie. Nourisson *9:* 65–86 (1921).

Marshall, J.A.: Deficient diets and experimental malocclusion considered from the clinical aspect. Int. J. Orthod. *18:* 438–449 (1932).

Marshall, W.A.: Problems in relating the presence of transverse lines in the radius to the occurrence of disease; in *Brothwell* The skeletal biology of earlier human populations. Symp. Soc. Study Human Biol., vol. 8, pp. 245–261 (Pergamon Press, Oxford 1968).

Massé, G.: Comparaison de la maturation osseuse chez de jeunes enfants de Dakar (Sénégal) et de Boston (Etats-Unis). Mod. Probl. Paediat., vol. 8, pp. 199–202 (Karger, Basel 1962).

Massé, G. and Hunt, E.E.: Skeletal maturation of the hand and wrist in West African children. Hum. Biol. *35:* 3–25 (1963).

Maun, M.E.; Cahill, W.M., and Davis, R.M.: Morphologic studies of rats deprived of essential amino acids. I. Phenylalanine. Archs Path. *39:* 294–300 (1945a).

Maun, M.E.; Cahill, W.M., and Davis, R.M.: Morphologic studies of rats deprived of essential amino acids. II. Leucine. Archs Path. *40:* 173–178 (1945b).

Maun, M.E.; Cahill, W.M., and Davis, R.M.: Morphologic studies of rats deprived of essential amino acids. III. Histidine. Archs Path. *41:* 25–31 (1946).

Mazess, R.B. and Cameron, J.R.: Skeletal growth in school children: maturation and bone mass. Am. J. phys. Anthrop. *35:* 399–407 (1971).

McCaffery, T.D.; Nasr, K.; Lawrence, A.M., and Kirsner, J.B.: Severe growth retardation in children with inflammatory bowel disease. Pediatrics *45:* 386–393 (1970).

McCance, R.A.: Some effects of undernutrition. J. Pediat. *65:* 1008–1014 (1964).

McCance, R.A.: The effect of normal development and growth of undernutrition on the growth of the calcified tissues. Tijdschr. soc. Geneesk. *44:* 569–573 (1966).

McCance, R.A.; Dickerson, J.W.T.; Bell, G.H., and Dunbar, O.: Severe undernutrition in growing and adult animals. Br. J. Nutr. *16:* 1–12 (1962).

McCance, R.A.; Ford, E.H.R., and Brown, W.A.B.: Severe undernutrition in growing and adult animals. 7. Development of the skull, jaws and teeth in pigs. Br. J. Nutr. *15:* 213–224 (1961).

McCance, R.A. and Widdowson, E.M.: Nutrition and growth. Proc. R. Soc. B *156:* 326–337 (1962).

McCay, C.M.; Crowell, M.F., and Maynard, L.A.: The effect of retarded growth upon the length of life span and upon the ultimate body size. J. Nutr. *10:* 63–79 (1935).

McCay, C.M.; Maynard, L.A.; Sperling, G., and Barnes, L.L.: Retarded growth, life span, ultimate body size and age changes in the albino rat after feeding diets restricted in calories. J. Nutr. *18:* 1–13 (1939).

McCrae, W.M. and Sweet, E.M.: Diagnosis of osteoporosis in childhood. Br. J. Radiol. *40:* 104–107 (1967).

McFie, J. and Welbourn, H.F.: Effect of malnutrition in infancy on the development of bone, muscle and fat. J. Nutr. *76:* 97–105 (1962).

McMeekan, C.P.: Growth and development in the pig, with special reference to carcass quality characters. II. The influence of the plane of nutrition on growth and development. J. agric. Sci. *30:* 387–436 (1940).

Morgan, D.B.; Paterson, C.R.; Woods, C.G.; Pulvertaft, C.N., and Fourman, P.: Search for osteomalacia in 1,228 patients after gastrectomy and other operations on the stomach. Lancet *ii:* 1085–1088 (1965a).

Morgan, D.B.; Paterson, C.R.; Woods, C.G.; Pulvertaft, C.N., and Fourman, P.: Osteomalacia after gastrectomy. Lancet *ii:* 1089–1091 (1965b).

Nawaby, M. El; Safwat Shukry, A.; Hefny, A. El., and Ragab, M.: A radiological study of the heart and bones in kwashiorkor in Egyptian children. J. Ind. pediat. Soc. *1:* 203–208 (1962).

Newberne, P.M.; Savage, J.E., and O'Dell, B.L.: Pathology of arginine deficiency in the chick. J. Nutr. *72:* 347–352 (1960).

Odland, L.M.; Mason, R.L., and Alexeff, A.I.: Bone density and dietary findings of 409 Tennessee subjects. I. Bone density considerations. Am. J. clin. Nutr. *25:* 905–907 (1972).

Ohlmüller, W.: Über die Abnahme der einzelnen Organe bei an Atrophie gestorbenen Kindern. Z. Biol. *18:* 78–103 (1882).

Outhouse, J. and Mendel, L.B.: The rate of growth. I. Its influence on the skeletal development of the albino rat. J. exp. Zool. *64:* 257–285 (1933).

Owens, P.D.A.: The effects of undernutrition and rehabilitation on the jaws and teeth of pigs; in *McCance and Widdowson* Calorie deficiencies and protein deficiencies, pp. 341–346 (Little, Brown, Boston 1968).

Park, E.A.: Bone growth in health and disease. Archs Dis. Childh. *29:* 269–281 (1954).

Park, E.A.: The imprinting of nutritional disturbances on the growing bone. Pediatrics *33:* 815–862 (1964).

Park, E.A. and Richter, C.P.: Transverse lines in bone: the mechanism of their development. Johns Hopkins Hosp. Bull. *93:* 234–248 (1953).

Parker, G.S.; Dreizen, S., and Spies, T.D.: A cephalometric study of children presenting clinical signs of malnutrition. Angle Orthod. *22:* 125–136 (1952).

Pickins, M.; Anderson, W.E., and Smith, A.H.: The composition of gains made by rats on diets promoting different rates of gain. J. Nutr. *20:* 351–365 (1940).

Platt, B.S. and Stewart, R.J.C.: Transverse trabeculae and osteoporosis in bones in experimental protein-calorie deficiency. Br. J. Nutr. *16:* 483–495 (1962).

Platt, B.S. and Stewart, R.J.C.: Effects of protein-calorie deficiency on dogs. I.,Reproduction, growth, and behaviour. Dvl Med. Child Neurol. *10:* 3–24 (1968).

Platt, H.S.; Stewart, R.J.C., and Platt, B.S.: Transverse trabeculae in the bones of malnourished children. Proc. Nutr. Soc. *22:* xxix–xxx (1963).

Pomeroy, R.W.: The effect of a submaintenance diet on the composition of the pig. J. agric. Sci. *31:* 50–73 (1941).

Posener, K.; Walker, E., and Weddel, G.: Radiographic studies of metacarpal bones in children. J. Anat. *74:* 76–79 (1939).

Poznanski, A.K.; Garn, S.M.; Kuhns, L.R., and Sandusky, S.T.: Dysharmonic maturation of the hand in the congenital malformation syndromes. Am. J. phys. Anthrop. *35:* 417–432 (1971).

Prader, A.; Shmerling, D.H.; Zachmann, M., and Biro, Z.: Catch-up growth in celiac disease. Acta pediat. scand. *58:* 311 (1969).

Prader, A.; Tanner, J.M., and Harnack, G.A. von: Catch-up growth following illness or starvation. J. Pediat. *62:* 646–659 (1963).

Pratt, C.W.M. and McCance, R.A.: Severe undernutrition in growing and adult animals. 2. Changes in the long bones of growing cockerels held at fixed weights by undernutrition. Br. J. Nutr. *14:* 75–89 (1960).

Pratt, C.W.M. and McCance, R.A.: Severe undernutrition in growing and adult animals. 6. Changes in the long bones during the rehabilitation of cockerels. Br. J. Nutr. *15:* 121–129 (1961).

Pratt, C.W.M. and McCance, R.A.: Severe undernutrition in growing and adult animals. 12. The extremities of the long bones in pigs. Br. J. Nutr. *18:* 393–408 (1964a).

Pratt, C.W.M. and McCance, R.A.: Severe undernutrition in growing and adult animals. 14. The shafts of the long bones in pigs. Br. J. Nutr. *18:* 613–629 (1964b).

Pryor, J.W.: The hereditary nature of variation in the ossification of bones. Anat. Rec. *1:* 84–88 (1907).

Pyle, S.I.; Mann, A.W.; Dreizen, S.; Kelly, H.J.; Macy, I.G., and Spies, T.D.: A substitute for skeletal age (Todd) for clinical use: the red graph method. J. Pediat. *32:* 125–136 (1948).

Quimby, F.H.; Bartlett, R.G., and Artress, J.L.: Effects of hormone, vitamin and liver supplements on the appetite and growth of the young rat during recovery from chronic starvation. Am. J. Physiol. *166:* 566–571 (1951).

Quinn, E.J.; King, C.G., and Dimit, B.H.: A study of the effects of certain diets upon the growth and form of albino rats. J. Nutr. *2:* 7–18 (1929).

Ramos Galvan, R.; Mariscal, A.C.; Viniegra, C.A. y Perez Ortiz, B.: Desnutrición en el niño (Impresiones Modernas, Mexico 1969).

Reddy, G.S.; Sastry, J.G., and Narasinga Rao, B.S.: Radiographic photodensitometric assessment of bone density changes in rats and rabbits subjected to nutritional stresses. Indian J. med. Res. *60:* 1807–1815 (1972).

Reichman, P. and Stein, H.: Radiological features noted on plain radiographs in malnutrition in African children. Br. J. Radiol. *41:* 296–299 (1968).

Restrepo, A.C.; Tejada, V.C. y Braham, E.: Estudio patológico de huesos en niños mal nutridos. Ant. med. *14:* 301–320 (1964).

Rey, J.; Rey, F.; Jos, J. et Amusquivar Lora, S.: Etude de la croissance dans 50 cas de maladie cœliaque de l'enfant. Archs fr. Pédiat. *28:* 37–47 (1971).

Reynolds, E.L.: Degree of kinship and pattern of ossification. Am. J. phys. Anthrop. *1:* 405–416 (1943).

Reynolds, E.L.: Differential tissue growth in the leg during childhood. Child Dev. *15:* 181–205 (1944).

Riesenfeld, A.: The effect of extreme temperatures and starvation on the body proportions of the rat. Am. J. phys. Anthrop. *39:* 427–460 (1973).

Riopelle, A.J.; Hale, P.A., and Watts, E.S.: Protein deprivation in primates. VII. Determinants of size and skeletal maturity at birth in rhesus monkeys. Hum. Biol. *48:* 203–222 (1976).

Roberts, L.J. and MacNair, V.: Effect of a milk supplement on bone development in institution children as indicated by ossification of bones of the wrist. J. Nutr. *13:* /9–10 (1937).

Roche, A.F.: Bone growth and maturation; in *Falkner and Tanner* Human growth: a comprehensive treatise (Plenum Publishing, New York, in press).

Roche, A.F. and Sunderland, S.: Multiple ossification centres in the epiphyses of the long bones of the human hand and foot. J. Bone Jt Surg. *41B:* 375–383 (1959).

Roche, A.F.; Wainer, H., and Thissen, D.: Skeletal maturity. The knee joint as a biological indicator. (Plenum Publishing, New York 1975).

Rohmann, C.G.; Garn, S.M.; Guzman, M.A.; Flores, M.; Béhar, M., and Pao, E.: Osseous development of Guatemalan children on low-protein diets. Fed. Proc. Fed. Am. Socs exp. Biol. 23: 338 (1964).

Salomón, J.B.; Blanco, R.; Arroyave, G. y Canosa, C.: Efecto de la nutrición sobre la formación del hueso compacto en niños preescolares. Archos latinamer. Nutr. 20: 29–39 (1970).

Salomón, J.B.; Klein, R.E.; Guzman, M.A. y Canosa, C.: Efectos de la nutrición e infecciones sobre desarrollo óseo de niños en una área rural en Guatemala. Archos latinamer. Nutr. 22: 417–449 (1972).

Satgé, P.; Mattei, J.F. et Dan, V.: Avenir somatique des enfants atteints de kwashiorkor. Annls paediat. 17: 368–381 (1970).

Saville, P.D. and Lieber, C.S.: Increases in skeletal calcium and femur cortex thickness produced by undernutrition. J. Nutr. 99: 141–144 (1969).

Saxton, J.A. and Silberberg, M.: Skeletal growth and ageing in rats receiving complete or restricted diets. Am. J. Anat. 81: 445–475 (1947).

Schneider, M. and Adar, U.: Effect of inanition of rabbit growth cartilage plates. Archs Path. 78: 149–156 (1964).

Schraer, H. and Newman, M.T.: Quantitative roentgenography of skeletal mineralization in malnourished Quechua Indian boys. Science 128: 476–477 (1958).

Schwartz, C.; Scott, E.B., and Ferguson, R.L.: Histopathology of amino acid deficiencies. I. Phenylalanine. Anat. Rec. 110: 313–327 (1951).

Scott, E.B.: Histopathology of amino acid deficiencies. III. Histidine. Archs Path. 58: 129–141 (1954).

Scott, E.B.: Histopathology of amino acid deficiencies. IV. Tryptophan. Am. J. Path. 31: 1111–1118 (1955).

Scott, E.B. and Schwartz, C.: Histopathology of amino acid deficiencies. II. Threonine. Proc. Soc. exp. Biol. Med. 84: 271–276 (1953).

Sedlmair, A.C.: Über die Abnahme der Organe, insbesondere der Knochen, beim Hunger. Z. Biol. 37: 25–48 (1899).

Shenolikar, I.S.: Concentrations of some elements in the human skeleton. Indian J. med. Res. 54: 1131–1137 (1966).

Shenolikar, I.S. and Rao, B.S.N.: Influence of dietary protein on calcium metabolism in young rats. Indian J. med. Res. 56: 1412–1422 (1968).

Shmerling, D.H.; Prader, A., and Zachmann, M.: The effect of dietary treatment on growth in coeliac disease; in McCance and Widdowson Calorie deficiencies and protein deficiencies, pp. 159–161 (Little, Brown, Boston 1968).

Shrader, R.E. and Zeman, F.J.: Effect of maternal protein restriction on postnatal growth hormone production and bone development in the rat. Fed. Proc. Fed. Am. Socs exp. Biol. 31: 679 (1972).

Shrader, R.E. and Zeman, F.J.: Skeletal development in rats as affected by maternal protein deprivation and postnatal food supply. J. Nutr. 103: 792–801 (1973).

Silberberg, M. and Silberberg, R.: Changes in cartilage and bone of immature female guinea pigs due to undernourishment. Archs Path. 30: 675–688 (1940).

Silberberg, R. and Silberberg, M.: Changes in bones and joints of underfed mice bearing anterior hypophyseal grafts. Endocrinology 60: 67–75 (1957).

Smith, D.M.; Nance, W.E.; Kang, K.W.; Christian, J.C., and Johnston, C.C.: Genetic factors in determining bone mass. J. clin. Invest. 52: 2800–2808 (1973).

Snodgrasse, R.M.; Dreizen, S.; Currie, C.; Parker, G.S., and Spies, T.D.: The association between anomalous ossification centers in the hand skeleton, nutritional status and rate of skeletal maturation in children five to fourteen years of age. Am. J. Roentg. *74:* 1037–1048 (1955).

Sobel, E.H.; Silverman, F.N., and Lee, C.M.: Chronic regional enteritis and growth retardation. Am. J. Dis. Child. *103:* 569–576 (1962).

Sontag, L.W. and Wines, J.: Relation of mothers' diets to status of their infants at birth and in infancy. Am. J. Obstet. Gynec. *54:* 994–1003 (1947).

Spies, T.D.; Dreizen, S.; Snodgrasse, R.M.; Arnett, C.M., and Webb-Peploe, H.: Effect of dietary supplement of nonfat milk on human growth failure. Am. J. Dis. Child. *98:* 187–197 (1959).

Spies, T.D.; Dreizen, S.; Snodgrasse, R.M.; Parker, G.S., and Currie, C.: Skeletal maturational progress of children with chronic nutritive failure. Am. J. Dis. Child. *85:* 1–12 (1953).

Stare, F.J.: Nutritional conditions in Holland. Nutr. Rev. *3:* 225–227 (1945).

Stewart, C.A.: Growth of the body and of the various organs of young albino rats after inanition for various periods. Biol. Bull. *31:* 16–51 (1916).

Stewart, C.A.: Changes in the relative weights of the various parts, systems and organs of young albino rats underfed for various periods. J. exp. Zool. *25:* 301–353 (1918).

Stewart, C.A.: Changes in the weights of the various parts, systems and organs in albino rats kept at birth weight by underfeeding for various periods. Am. J. Physiol. *48:* 67–78 (1919).

Stewart, R.J.C.: Bone pathology in experimental malnutrition. Wld Rev. Nutr. Diet., vol. 21, pp. 1–74 (Karger, Basel 1975).

Stewart, R.J.C. and Platt, B.S.: Arrested growth lines in the bones of pigs on low-protein diets. Proc. Nutr. Soc. *17:* v-vi (1958).

Stini, W.A.: Nutritional stress and growth: sex differences in adaptive response. Am. J. phys. Anthrop. *31:* 417–426 (1969).

Suckling, P.V. and Campbell, J.A.: A five-year follow-up of coloured children with kwashiorkor in Cape Town. J. trop. Pediat. *2:* 173–180 (1957).

Tanner, J.M.; Whitehouse, R.H., and Healy, M.J.R.: A new system for estimating skeletal maturity from the hand and wrist, with standards derived from a study of 2,600 healthy British children (International Children's Centre, Paris 1962).

Thurm, D.A.; Samonds, K.W., and Hegsted, D.M.: The effects of a 20-week nutritional insult on the skeletal development of *Cebus albifrons* during the first year of life. Am. J. clin. Nutr. *29:* 621–625 (1976).

Todd, T.W.: Differential skeletal maturation in relation to sex, race, variability and disease. Child Dev. *2:* 49–65 (1931).

Todd, T.W.: Atlas of skeletal maturation, p. 203 (Mosby, St. Louis 1937).

Tompkins, W.T. and Wiehl, D.G.: Epiphyseal maturation in the newborn as related to maternal nutritional status. Am. J. Obstet. Gynec. *68:* 1366–1377 (1954).

Tonge, C.H. and McCance, R.A.: Severe undernutrition in growing and adult animals. 15. The mouth, jaws and teeth of pigs. Br. J. Nutr. *19:* 361–372 (1965).

Tonge, C.H. and McCance, R.A.: Normal development of the jaws and teeth in pigs, and the delay and malocclusion produced by calorie deficiencies. J. Anat. *115:* 1–22 (1973).

Toverud, K.U. and Toverud, G.: Chemical and histological studies of bones and teeth of newborn infants. Acta paediat., Uppsala *16:* 459–467 (1933).

Trowell, H.C.; Davis, J.N.P., and Dean, R.F.A.: Kwashiorkor (Arnold, London 1954).

Variot, M.G.: Sur les caractères spéciaux de l'hypotrophie chez les prématurés. Atrophie pondérale. Atrophie staturale. Soc. Pédiat., Paris *9:* 363–369 (1907).

Vaughan, J.M.: The physiology of bone (Clarendon Press, Oxford 1970).

Voit, C.: Über die Verschiedenheiten der Eiweisszersetzung beim Hungern. Z. Biol. *2:* 308–365 (1866).

Walker, A.R.P. and Arvidsson, U.B.: Studies on human bone from South African Bantu subjects. I. Chemical composition of ribs from subjects habituated to a diet low in calcium. Metabolism *3:* 385–391 (1954).

Wallace, L.R.: The growth of lambs before and after birth in relation to the level of nutrition. J. agric. Sci. *38:* 367–401 (1948).

Warkany, J. and Nelson, R.C.: Appearance of skeletal abnormalities in the offspring of rats reared on a deficient diet. Science *92:* 383–384 (1940).

Warkany, J. and Nelson, R.C.: Skeletal abnormalities in the offspring of rats reared on deficient diets. Anat. Rec. *79:* 83–100 (1941).

Waterlow, J.C. and Alleyne, G.A.O.: Protein malnutrition in children: advances in knowledge in the last ten years. Adv. Protein Chem. *25:* 117–241 (1971).

Weiske, H.: Über den Einfluss der Nahrungsentziehung auf das Gewicht und die Zusammensetzung der Organe, insbesondere der Knochen und Zahne. Hoppe-Seyler's Z. physiol. Chem. *22:* 485–499 (1897).

Wellman, O.: Untersuchungen über den Umsatz von Ca, Mg und P bei hungernden Tieren. Pflügers Arch. ges. Physiol. *121:* 508–533 (1908).

WHO: Nutrition: a review of the WHO programme. 1. Wld Hlth Org. Chron. *26:* 160–179 (1972).

Widdowson, E.M.: The effect of growth retardation on postnatal development; in *Lodge and Lamming* Growth and development of mammals, pp. 224–233 (Plenum Publishing, New York 1968).

Widdowson, E.M.: Intra-uterine growth retardation in the pig. Biol. Neonate *19:* 329–340 (1971).

Widdowson, E.M.: Changes in pigs due to undernutrition before birth, and for one, two and three years afterwards, and the effects of rehabilitation; in *Roche and Falkner* Nutrition and malnutrition, identification and measurement. Advances in experimental medicine and biology, vol. 49, pp. 165–182 (Plenum Publishing, New York 1974).

Widdowson, E.M. and Dickerson, J.W.T.: Chemical composition of the body; in *Comar and Bronner* Mineral metabolism. An advanced treatise, vol. 2, pp. 1–247 (Academic Press, New York 1964).

Widdowson, E.M. and McCance, R.A.: The effect of finite periods of undernutrition at different ages on the composition and subsequent development of the rat. Proc. R. Soc. B *158:* 329–342 (1963).

Winters, J.C.; Smith, A.H., and Mendel, L.B.: The effects of dietary deficiencies on the growth of certain body systems and organs. Am. J. Physiol. *80:* 576–593 (1927).

Yarbrough, C.; Habicht, J.-P.; Klein, R.E., and Roche, A.F.: Determining the biological age of the preschool child from a hand-wrist radiograph. Investve Radiol. *8:* 233–243 (1973).

Yarbrough, C.; Martorell, R.; Klein, R.E.; Himes, J.H.; Malina, R.M., and Habicht, J.-P.: Stature and age as factors in the growth of second metacarpal cortical bone in moderately malnourished children. Ann. hum. Biol. *4:* 43–48 (1977).

Yeager, R. and Winters, J.C.: The effect of deficient diets on the total ash, calcium and phosphorus content of bones. J. Nutr. *10:* 389–397 (1935).

Zayaz, S.L.; Mack, P.B.; Sprague, P.K., and Bauman, A.W.: Nutritional status of school children in a small industrial city. Child Dev. *11:* 1–25 (1940).

Zucker, L.M. and Zucker, T.F.: Specific and non-specific nutritional effects as illustrated in bone. Am. J. Physiol. *146:* 593–599 (1946).

Dr. *J.H. Himes,* Fels Research Institute, *Yellow Springs, OH 45387* (USA)

Wld Rev. Nutr. Diet., vol. 28, pp. 188–209 (Karger, Basel 1978)

Hepatocarcinogens in Nigerian Foodstuffs

Enitan A. Bababunmi, Anthony O. Uwaifo and Olumbe Bassir

The Department of Biochemistry, University of Ibadan, Ibadan, Oyo State

Contents

I. Introduction

Nigeria is Africa's most populous country in spite of its relatively small area of 356,669 square miles. The population of Nigeria is estimated to be well over

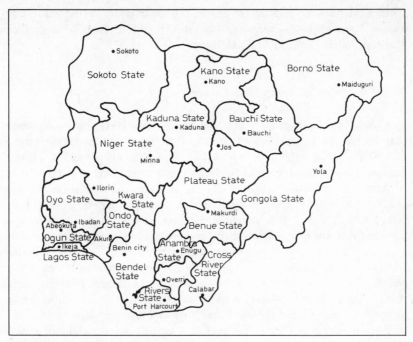

Fig. 1. The 19 United States of Nigeria.

70 million. Although the proportional contribution of agriculture to Nigerian economy continues to diminish, its dominance as the single most important sector in the economy will remain for a very long time (*Aboyade,* 1971).

It is a curious fact that many peasant communities feed on plant and animal tissues which often contain, before processing, chemicals which are toxic to the living cell. Some of these foodstuffs are rich in nutrients, vitamins, amino acids and trace elements which make them particularly desirable. By trial and error, through the ages, peasants learn the best culinary processes by which such food toxicants can be got rid of. Of particular interest to Third World nutritionists are mycotoxins, such as the aflatoxins, which are metabolites of certain strains of fungi that contaminate food accidentally as a result of environmental pollution. Some of these toxins have been shown to be lethal in minute doses, while others appear to have beneficial or therapeutic properties when ingested in sub-nanogram quantities. Modern food technology carries its own hazards. Preservatives, colouring matter and emulsifiers are just a few possible food toxicants.

Nitrosamines emanating from nitrites and pyrrolizidine alkaloids in plants which are used as medicinal herbs present a serious risk. The modes of action of these substances, as carcinogenic agents, and the implications in the aetiology of human hepatocellular carcinoma are still posing a formidable intellectual challenge to biochemists and toxicologists.

II. Incidence of Human Liver Cancer

The incidence of primary liver cancer in Nigeria is very high compared with that in Europe or the United States of America. The incidence rates (per 100,000) in adult males were reported by *Doll et al.* (1966) as 5.9, 2.0 and 1.8 for Ibadan (Nigeria), Liverpool (United Kingdom) and New York (United States of America), respectively. Ibadan city has an estimated population of well over one million. It lies in the forest belt and is the largest city in West Africa. In a survey (*Odebiyi*, 1972), it was observed that about 200 cancer patients were admitted into the Ibadan University Teaching Hospital yearly; carcinomas of the liver, stomach and oesophagus were most common. The relative ratio frequency of hepatocellular carcinoma in adult male Nigerians is relatively high (*Edington and MacLean*, 1965; *Hutt*, 1971). However, the occurrence of hepatic neoplasm cases in the children in Ibadan is quite low (*Williams et al.*, 1967). With increase in age, *Williams* (1975a) observed that there is a relatively steep increase in the number of diagnosed cases of liver cell carcinoma. In addition, malignant tumours are diagnosed in Nigeria in more advanced stages than in industrialized countries.

Hutt (1971) stated that the high-incidence geographical areas are characterized by a shift of the age-incidence profile towards the younger age groups. An early exposure to carcinogenic agents has been suggested as a possible cause for this shift. Some of these cancer-inducing agents are undoubtedly present in foods. Indeed, a major way by which nutrition could influence cancer incidence is the initiation, promotion or acceleration of the carcinogenic process by nutrients, food additives or contaminants. This aspect of cancer research is beginning to receive some attention (*Wynder*, 1975).

III. Common Nigerian Foodstuffs

Nigeria stretches across a wide range of ecological zones and it is inhabited by a population of heterogenous cultural background. Consequently, there is a diversity of foodstuffs which are produced and consumed by the people (*Oyenuga*, 1968). The pattern of food in a country is determined to a large extent by the ecology within its geographical confines and by the movements of human population and foodstuffs across its political boundaries (*Idusogie*, 1973). The foodstuffs which are consumed in Nigeria can be classified as

Table I. Projection of demand for foodstuffs in Nigeria, 1968–1985[1]

Foodstuffs	Foodstuffs, grams per capita per day	
	1968 (population 62,985,000)	1985 (population 102,456,000)
Maize	36.187	38.064
Millet	83.055	87.362
Sorghum	129.857	136.589
Rice	14.527	18.451
Wheat	1.206	2.187
Acha	1.200	1.260
Cassava	327.690	272.564
Sweet potatoes	5.929	4.932
Irish potatoes	0.608	0.504
Yam	315.377	262.322
Cocoyam	34.946	29.069
Plantain	54.482	45.317
Groundnuts	11.464	10.787
Cowpeas	18.764	17.659
Soyabeans	1.696	1.595
Bambara nuts	1.581	1.488
Vegetables	50.715	64.316
Beef	0.907	1.548
Goat meat	5.042	8.058
Mutton	1.156	1.850
Poultry	2.299	4.428
Pork	1.321	1.677
Bush meat	10.149	14.464
Fish	36.658	62.488
Beer	3.847	4.885
Milk	7.568	12.922
Fruits	5.836	8.042
Palm oil	23.350	27.074
Mellon seed oil	0.389	0.452

[1] From *Olayide et al.* (1972).

follows: (a) legumes; (b) cereals; (c) roots, tubers and starchy fruits; (d) nuts and oil seeds; (e) vegetables; (f) fruits; (g) meat, poultry and fish, and (h) beverages.

The types and amounts of foods which are preferred and eaten by a Nigerian are normally determined by the individual's ethnic origin, education and social class. In general, roots, tubers and starchy fruits are the predominant source of food for the majority of the population.

A. Legumes

In common with other tropical, sub-tropical and temperate regions of the world, a variety of legumes (*Aykroyd and Joyce*, 1964) are consumed by Nigerians. These include cowpea *(Vigna anguiculata)*, lentils *(Lens esculenta)*, pigeon pea *(Cajanus caja)*, soya beans, groundnuts *(Voandzeia subterranea)*, African locust bean *(Parikia africana)*, African walnut, African oil bean *(Pentaclethra macrophylla)* and the horse eye bean *(Mucuna soloanei)*. Data on the projection of demand for foodstuffs in Nigeria for the period of years 1968 to 1985 (in grams per capita per day) are shown in table I. These data represent accurate indices for food consumed in 1968 and for that which would be consumed in 1985.

B. Cereals

The principal cereals which are eaten in Nigeria and most other African countries are foreign to the continent. They include wheat *(Triticum vulgare)*, rice *(Oryza sativa)*, barley *(Hordium vulgare)*, sorghum (*Sorghum* spp.), millet (*Pennisetum typhoides* and *Elensine coracana*) and maize *(Sea mays)*. Data on the amount of some of these cereals (in grams per capita per day) in demand for food in the years 1968 and 1985 are also given in table I.

C. Roots, Tubers and Starchy Fruits

Nigerians, like other Africans who inhabit the west coast of Africa, are traditionally 'root-eaters' (*Idusogie*, 1973). Yam (*Dioscorea* spp.) is one of the food crops which come under this group. It is the traditional staple food of many Nigerians. The varieties eaten are: white yam *(Dioscorea rotundata)*, Yellow yam *(Dioscorea cayenensis)*, water yam *(Dioscorea alata)* and bitter yam *(Dioscorea dumetrorum)*. The importance of yam in the nutrition of Nigerians is reflected in the fact that Nigeria produces about half the world's yam crop (*Oyenuga*, 1968; *Coursey and Haynes*, 1970). Cassava *(Manihot esculenta)* is another important member of this food group. It is also widely consumed in Nigeria. Other food groups in the group include: cocoyam (*Calocasia* spp.), sweet potato *(Ipomoea batatas)* and Irish potato *(Solnum tuberosum)*. Starchy fruits of plants, such as boabab tree *(Adansonia digitata)*, African breadfruit *(Treculia africana)*, the breadfruit tree *(Artrocarpus altilis)*, plantain *(Musa paradisiaca)* and banana *(Musa sapientum)*, are also widely consumed.

D. Nuts and Oil Seeds

Relatively few indigenous nuts and oil seeds are consumed in Nigeria. The common ones are: kolanuts *(Cola acuminata)*, bitter cola *(Dennettia tripetala)*, bush mango or dikanut *(Irvingia gabnensis)*, the oil palm *(Elaeis guinensis)*, melon seeds *(Citrillus vulgarie)*, water melon *(Citrillus lanatus)* and coconuts *(Cocoa mucifera)*.

Table II. Nigeria's major farm-fresh edible leafy vegetables

Local name	Botanical name
Soko	*Celocia argentea*
Tete	*Amaranthus hybridus*
Gbure	*Talinum triangulare*
Ewedu	*Corchorus olithorus*
Ewuro	*Vernonia amygadalina*
Igbo	*Solanum africana*
Ogunmo	*Solanum nodiflorum*
Ugu	*Telfairia occidentalis*
Uziza	*Piper guineense*
Ukazi	*Gnetum buchholzianum*
Ewe paki (cassava leaves)	*Manihot esculenta*
Ewe koko	*Xanthosoma sagittifolium*
Ilasa (okro leaves)	*Hibiscus esculentus*

E. Vegetables

Numerous species of leafy vegetables form part of the staple diets of all people in Nigeria. These vegetables are harvested at all stages of growth and are eaten at least once daily. They have a high crude protein content and are thus an important source of protein for numerous people (*Oke,* 1967; *Oke and Umoh,* 1974; *Fafunso and Bassir,* 1976). Table II shows the list of some farm-fresh edible leaves in Nigeria.

F. Fruits

Many varieties of fruits are consumed in Nigeria. Most of these fruits grow wildly while a few are cultivated in either orchards or near living homes. The varieties include common fruits such as: mango *(Magnifera indica),* oranges *(Citrus aurantium),* pawpaw *(Carcia papaya),* pineapple *(Ananas comoso),* grape *(Citrus paradisa),* lemon *(Citrus limon),* pepper fruits (*Dennettia* spp.) and African cherry *(Chrysophyllum albidum).* Some other less common fruits grow naturally in forests.

G. Meat, Poultry and Fish

Meat consumed by Nigerians is derived from a large collection of different types of animals. These animals live in all the ecological niches. The pool of animals includes: fish, birds of all kinds and sizes (chicken, turkey, guinea-fowl, pheasants, hawks, parrot, bats and toucan), elephant, hippopotamus, bush pig, chimpanzee, monkeys, antelope *(Tragelaphinae),* pangoline *(Manis longicandata),* porcupine *(Atherurus africanus),* wild dog, giant cambian rat *(Cricetomys gambianus),* grass cutter or cane rat *(Thryonomys swinderianus),* jungle or bush cat, monitor lizard, python and other snakes, African squirrel (*Paracerus* spp.), crocodile (*Crocodylis* spp.), snails, tortoise, turtle, anteater, periwinkles, frog, silkworms, termites, locust, and grasshoppers. There are other types of domesticated animals such as goats, sheep, cattle and ducks which are also eaten. The non-domesticated animals are usually captured by trapping or by hunting. Meat derived from this class constitutes a delicacy usually referred to as bush-meat.

H. Beverages

Both foreign and indigenous alcoholic beverages are consumed in Nigeria. These include imported and locally brewed beer, imported and indigenous liquor, imported wines and many indigenous wines. The indigenous wines which are consumed in different societies of the Nigerian population are: palm wine (fermented saps of *Elaeis guinensis* or *Raphia vinitera*), burukutu (extract of fermented mixture of malted *Sorghum vulgers* and *Manihot* flour), pito (fermented extract of malted maize and/or sorghum), otiagbagba (extract of fermented over-ripe *Musa sapientas*), ogogoro (crude alcoholic distillate of palm-wine), nono (defatted, slightly fermented fresh milk of cow), and cocoa wine (fermented aqueous extract of cacao seeds).

I. Miscellaneous

Apart from the use of plants as foods, many plant and animal products are consumed by Nigerians for curative purposes. These plants which are used as native medicines contain substances which can have therapeutic value but when ingested in sufficient quantities may be highly toxic. The plant species of *Senecio, Crotalaria* and *Clausena* which contain the pyrrolizidine and carbazole alkaloids have been identified as part of herbal teas in Nigeria (*Okorie,* 1975; IARC, 1976).

The crushed roots of the Crotalaria and sometimes its fresh juice are often used to relieve fever. *Clausena* roots are used in local medicine for the cure of headache, toothache and haemorrhoids. The bark of the *Cinchona* tree which is often used against malaria contains quinine (the oldest antimalarial medicine) and quinic acid. These two substances are known to be toxic in fairly large doses. Also, the leaves of some shrubs, such as the *Senecio* species, are extracted for the treatment of malaria, the deadliest parasitic disease in tropical Africa. The leaves of *Ocimium gratissimum* are extracted for treatment of dysentry, diarrhoea and other stomach upsets. Water extract of the leaves of sweet potato *(Ipomoea batatas)* when mixed with palm wine gives a mixture which is used to treat gonorrhea in the Bendel State of Nigeria.

Examples of plant foods which are presumed to have medicinal properties are: bitter leaf, bitter kola and kolanut. The leaves and roots of bitter leaf are used for the treatment of cough and malaria. The leaves are also used for prepar-

Table III. Fungi isolated from market-purchased gari[1]

Aspergillus candidus	*Cladosporium* sp.
Aspergillus flavus	*Curvularia* sp.
Aspergillus niger	*Fusarium solani*
Aspergillus tamarii	*Penicillium variable*
Aspergillus versicolor	*Rhizopus* sp.

[1] From *Adeniji* (1976).

ing the traditional sauce known as 'West African pallava sauce'. Bitter kolas are mixed with honey to give a preparation which is used for the treatment of cough. It is also believed that the consumption of bitter kola invigorates sexual activities.

IV. Occurrence of Carcinogens in Foods

The load of carcinogens in Nigerian foodstuffs can be determined by assessing: (a) the contamination of staple food commodities with fungal species; (b) the production of N-nitroso compounds in local beverages, and (c) the nutritional plant materials with definite therapeutic value for the presence of alkaloids and other related substances.

A. Fungal Contamination

Fungal spoilage and the consequent contamination of foods by microbial toxins has always been a potential danger in food production. In tropical and sub-tropical regions of the world, the risk is more ominous because of the typical tropical climate (high temperature and high relative humidity) which normally favour the growth of moulds in food (*Purchase,* 1968; *Purchase and van der Watt,* 1970).

The staple foods of the native people are usually fermented plant and animal products (*Akinrele,* 1964). During the preparation of *gari* (a staple and popular food among the low income group in the southern regions), fermentation of cassava *(Manihot utilissima)* is a very important step. At least ten different fungi were found by *Adeniji* (1976) to be associated with market and stored gari (table III). The *Aspergilli* and *Penicillia* species are among the fungi which were responsible for the deterioration and taste tainting of gari. No sample of market gari screened was found to be free of moulds. It has been recommended that, for long-term storage, gari should be kept at a relatively low moisture content less than 2%. When nine storage sites were sampled regularly in

Table IV. Fungi isolated from commercial cocoa in Nigeria[1]

Major pathogens	Minor pathogens
Aspergillus chevalieri	*Absidia corymbifera*
A. flavus	*Aspergillus aculeatus*
A. fumigatus	*A. nidulans*
A. niger	*A. ochraceus*
A. penicilloides	*A. pseudoglaucus*
A. ruber	*A. restrictus*
A. tamarii	*A. sydowi*
Botryodiplodia theobromae	*Curvularia lunata*
Mucor pusillus	*Cylindrocarpon tonkinense*
Paecilomyces varioti	*Fusarium equiseti*
Penicillium citrinum	*F. oxysporum*
Rhizopus arrhizus	*F. solani*
Syncephalastrum racemosum	*Geotricum candidum*
	Geotricum sp.
	Macrophoma sp.
	Penicillium decumbens
	P. Steckii
	P. variabile

[1] From *Oyeniran and Adeniji* (1975).

Ibadan, *Oyeniran and Adeniji* (1975) demonstrated that there was an extensive mould deterioration of cocoa during processing and storage. Toxigenic strains of *Aspergillus* and *Penicillium* (table IV) enter the cocoa beans at different stages, such as preharvest, fermentation, drying or storage. The degree of internal mouldiness in Nigeria cocoa has been found to be between 0.7 and 1.5%. These two fungal species have also been associated with storage decay of yam in Nigeria (*Adeniji*, 1970).

Aspergillus flavus, the main microbial source of aflatoxin (a potent liver carcinogen) is one of the common microbial contents of air and soil. It is found on living or dead plants and animals throughout the world (*Semeniuk*, 1964). In a favourable environment of relative humidity (70–99%) and a minimal temperature of about 10 °C, the fungus will grow on food materials (*Deiner and Davis*, 1968). In general, growth of *Aspergillus flavus* is correlated with the production of aflatoxin except at high temperatures of 40–50 °C. *Deiner and Davis* (1970) observed that aflatoxin formation in peanuts (groundnuts) was generally correlated with a kernel moisture content of 10% or higher. Groundnuts and groundnut oil are major export commodities and in this respect Nigeria supplied approximately 35% of the total world supply of groundnuts 10 years ago. In Nigeria, food storage facilities and preservation techniques are inadequate. The combination of this unfortunate situation, the natural warm and moist weather, squalid environment, human error and ignorance provides conducive atmosphere for the growth of *Aspergillus flavus* and consequently the elaboration of aflatoxin on foodstuffs.

In many African countries, assessment of aflatoxin contamination of diets showed a significant association with the relatively high incidence of human liver cancer in these countries (*Keen and Martin,* 1971; *Peers and Linsell,* 1973; *Peers et al.,* 1976). In the African town of Mozambique (a high human primary liver cancer area) the mean daily per capita ingestion of aflatoxin through foods was calculated to 15 µg per adult per day (*Van Rensburg et al.,* 1974; *Wogan,* 1975a). The principal food crops in Nigeria are yam, cassava, maize, rice, beans and millet. When common food preparations of these crops were sampled from local market stalls and assessed for aflatoxin contamination by the use of conventional techniques (AOAC, 1975), the aflatoxin content was not less than 0.5 ppm in any of the foods (*Bababunmi et al.,* 1976). Amounts of the order of 90 ppb of aflatoxin have also been found in locally hawked vegetables (*Bassir,* 1969) such as *Vernonia amygadalina* (bitter leaf). The *Penicillium* fungal species is also known to produce aflatoxin (*Hodges et al.,* 1964; *Ciegler et al.,* 1966). The brown root rot of tomato in *Colletotrichum* infection has partially been associated with *Penicillium* fungus (*Adeniji,* 1966).

B. Synthetic Reactions

Organic *N*-nitroso compounds, such as nitrosamines, occur either naturally in foods or as the product of reactions between nitrite and secondary amines in food. This class of compounds may also be formed from bacterial action. Nitrosamines present a serious health hazard. More than 80% of the nitroso compounds which have been tested have proven to be carcinogenic in experimental animals. Numerous reviews on the formation and occurrence of these substances have appeared during the last decade (*Magee and Barnes,* 1967; *Lijinsky and Epstein,* 1970; *Bogovski et al.,* 1972; *Fiddler,* 1975; *Mirvish,* 1975; *Issenberg,* 1976).

Although there have been difficulties in the search for nitrosamines in local foods, the recent arrival in Ibadan and the use of the techniques of gas chromatography and mass spectrophotometry have facilitated the detection and determination of these carcinogens in some beverages. The results of the survey (*Joaquim,* 1973) for nitrosamine in some popular Nigerian beverages indicated that there is a definite nitrosámine contamination of beverages in many areas (fig. 1) of the country. Tables V and VI give the distribution of the contamination of two types of *N*-nitroso compounds in major towns of the Southern states of Lagos, Ogun, Oyo and Ondo. The recent results of *Maduagwu* (1976) indicated that the hepatocarcinogenic dimethylnitrosamine occurs at the microgram per litre levels in palm-wine (the most popular alcoholic beverage) sampled

Table V. Concentration of nitrosamines in alcoholic beverages: Oyo, Ogun, Ondo and Lagos States of Nigeria[1]

Beverage	% of sample number containing nitrosamine	Dimethylamine ng per litre	Diethylamine ng/litre
Ogogoro	100	52	46
Burukutu	100	29	22
Pito	100	25	23
Oti-agbagba (agadagidi)	80	14	–

[1] From *Joaquim* (1973).

in 15 different towns of Kwara and Benue States (table VII). Palm wine, Oti-agbagba, nono, burukutu, pito and ogogoro are drunk by many Nigerians in fairly large volumes. It is estimated that a volume of about 2.5 litres of palm wine can be consumed at one sitting by a common man. Most Nigerians can afford the cost of this beverage, for their pleasure. In addition to the presence of dimethyl- and diethylnitrosamine in these beverages, there are speculations that other volatile and non-volatile nitroso compounds may be present in much smaller quantities. The least concentration of dimethylnitrosamine in ogogoro (a very popular locally distilled spirit) is of the order of 50 ng per litre.

Nitrosamines can be formed by bacterial action (*Ayanaba and Alexander,* 1973; *Issenberg,* 1976). There is some evidence that microorganisms in palm wine are involved in the formation of nitrosamines and that the rate of nitrosamine formation increases linearly with fermentation time (*Joaquim,* 1973).

C. Plant Constituents

Many plant species (*Dalziel,* 1948) grow and are used as food and native medicines in Nigeria. The plant general of *Crotolaria, Senecio, Laburnum* and *Heliotropium* have long been known to contain hepatocarcinogens (*Miller and Miller,* 1976). Many other plant foods contain several toxic substances which should be tested for carcinogenic properties. The pyrrolizidine alkaloids, monocrotaline and retrorsine which occur naturally in local plants, appear to be highly carcinogenic to experimental animals.

V. Biological Data

There have been several reviews on the effects of liver carcinogens on animals, plants and microorganisms (*Wogan,* 1966; *Legator,* 1969; IARC, 1972;

Table VI. Concentration of nitrosamines in stale palm-wine samples from different towns: Oyo, Ogun, Ondo and Lagos States of Nigeria[1]

Towns	% of sample number containing nitrosamine	Dimethylnitrosamine ng per litre	Diethylnitrosamine ng per litre
Abeokuta	100	15	11
Akure	100	18	13
Badagry	100	17	13
Epe	100	16	12
Ibadan	100	17	12
Ife	100	16	12
Ljebu-Ode	95	16	12
Ikorodu	95	17	12
Ogbomosho	95	17	12
Okitipupa	100	17	13
Ondo	100	18	14
Oshogbo	100	17	12
Otta	100	17	13
Oyo	95	18	12
Shagamu	100	17	11

[1] From *Joaquim* (1973).

Table VII. Dimethylnitrosamine contamination of palm-wine in Kwara and Benue States of Nigeria[1]

Sampling area	Concentration of dimethylnitrosamine in palm-wine, μg per litre ± SD
Ankpa[2]	6.7 ± 4.5
Dekina[2]	15.1 ± 4.3
Egbe	7.8 ± 3.9
Idah[2]	12.3 ± 8.0
Ilorin	13.1 ± 6.5
Jebba	13.1 ± 9.9
Kabba	4.4 ± 8.1
Kaima	22.3 ± 10.5
Lokoja	7.5 ± 4.9
Luma	11.3 ± 9.1
New-Bussa	8.9 ± 7.2
Offa	8.1 ± 6.5
Okene	8.3 ± 5.8
Omu-Aran	26.5 ± 11.4
Pateyi/Lafiagi	4.7 ± 2.6

[1] From *Maduagwu* (1976); at least 20 samples were tested in each area.
[2] Benue State areas. Others are situated in Kwara State.

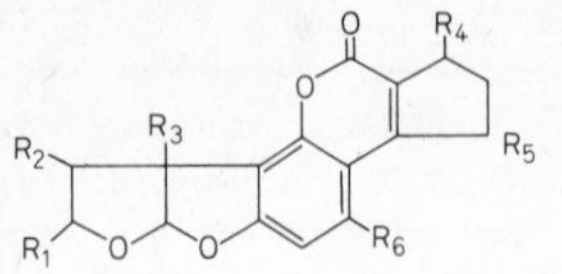

Aflatoxin	R_1	R_2	R_3	R_4	R_5	R_6
B_1	H	H	H	=O	H	OCH_3
B_2	H_2	H_2	H	=O	H	OCH_3
B_{2a}	HOH	H_2	H	=O	H	OCH_3
M_1	H	H	OH	=O	H	OCH_3
M_2	H_2	H_2	OH	=O	H	OCH_3
P_1	H	H	H	=O	H	OH
Q_1	H	H	H	=O	OH	OCH_3
R_o	H	H	H	OH	H	OCH_3

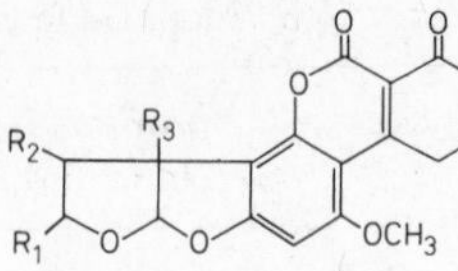

Aflatoxin	R_1	R_2	R_3
G_1	H	H	H
G_2	H_2	H_2	H
G_{2a}	OH	H_2	H
GM_1	H	H	OH

Fig. 2. The family of aflatoxin B_1.
Fig. 3. The family of aflatoxin G_1.

Miller and Miller, 1976). Different species of animal have been used in the demonstration of hepatic neoplasia induced by various carcinogens. Although it is not possible to draw definite and direct conclusions from animal experimentation, the following biological data of four different carcinogens in foods are relevant in the assessment of the carcinogenic risk to the Nigerian. The risk to man of each of these substances has been assessed adequately in previous reviews (IARC, 1976).

A. Aflatoxin

Since the discovery and isolation of aflatoxin about 15 years ago, this hepatocarcinogen has been assayed in a wide variety of biological systems more than any other known mycotoxin (*Legator,* 1969). The earliest indications that a groundnut meal in the tropics could cause a liver disease was given by *Asplin and Carnaghan* (1961) and *Blount* (1961). This poisonous substance is a metabolite of *Aspergillus flavus* which is a bifurano-coumarin compound. Aflatoxin B_1 (fig. 2) is most frequently present in mould-contaminated foodstuffs. Figure 3 shows a series of other types of aflatoxin (*Wogan,* 1975a). Table VIII shows some results of lethal dose (LD_{50}) studies for aflatoxin B_1. Since the aflatoxin molecule possesses the α, β-unsaturated lactone ring, its biological activity could be related to its high reactivity with electrophilic agents and its potential action as an alkylating agent under physiological conditions (*Dickens and Jones,* 1963; *Dickens,* 1964).

Dietary aflatoxin B_1 has been shown to be carcinogenic (IARC, 1976) in several species of animal (for example, duck, rat and rainbow trout) at ppb levels. Oral administration of aflatoxin to rhesus monkeys induces hepatic

Table VIII. Lethality of single oral doses of aflatoxin B_1 in some animals[1]

Animal	Age	Sex	LD_{50} mg per kg
Duckling	1 day	M	0.37
Rat	21 days	M	5.5
	21 days	F	7.4
Hamster	30 days	M	10.2
Dog	adult	M/F	0.5 approx.

[1] From *Wogan* (1966).

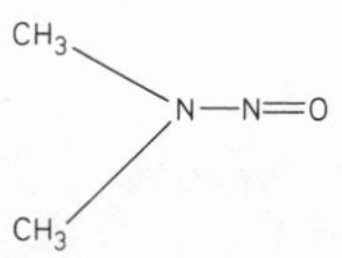

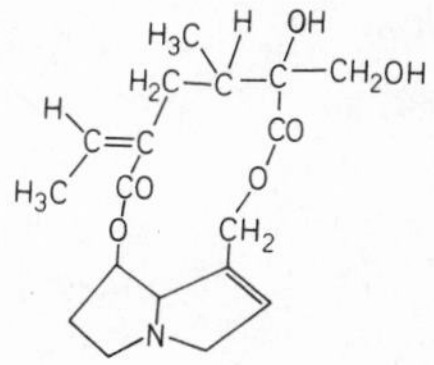

Fig. 4. Dimethylnitrosamine.
Fig. 5. Monocrotaline.
Fig. 6. Retrorsine.

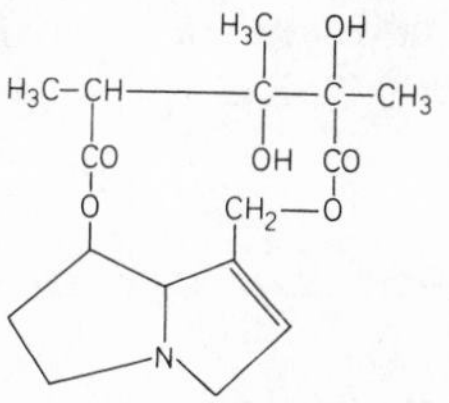

fibrosis in these non-human primates (*Madhavan et al.,* 1965). In relation to man, incubation of aflatoxin B_1 with a human liver (or rat liver) microsomal system in the presence of *Salmonella typhimurium* produces reverse mutations in the bacteria (*Ames et al.,* 1973a; *McCann et al.,* 1975). There is a significant degree of correlation between mutagenicity and carcinogenicity (*Rohrborn,* 1974).

B. Nitrosamine

The most common carcinogenic *N*-nitroso compound in Nigerian foods (or beverages) is *N*-nitrosodimethylamine (dimethylamine, DMN). DMN was identified by *Ender et al.* (1964) as the cause of liver disorders in domestic animals which fed on nitrite-treated meals. In general, the nitrosamines are potent hepato-carcinogens. Comprehensive reviews on the animal toxicology of this class of compounds have been written by *Magee* (1971), *Wolff and Wasserman* (1972) and *Shank* (1975).

The report by IARC (1972) states that DMN (fig. 4) induces tumours (*Schoental*, 1976) in at least seven animal species, namely mouse, rat, hamster, guinea-pig, rabbit, rainbow trout and mastomys. The target organs are the liver and the kidney. The carcinogenic compound appears to induce neoplasm transplacentally (*Shank*, 1975).

C. Monocrotaline

Monocrotaline is a pyrrolizidine alkaloid (fig. 5) of *Crotalaria* plant origin. Following oral administration, it is carcinogenic in rats and produces liver carcinoma. Some liver lesions which are observed in calves are very similar to those in man who had suffered from *Senecio* plant poisoning (*Bras et al.*, 1957).

D. Retrorsine

This substance (fig. 6) is also carcinogenic and produces a variety of tumours in rats when it is administered orally (IARC, 1976). Retrorsine is one of the main alkaloids present in the *Senecio* plant species.

VI. Development of Hepatocellular Carcinoma in Man

20 years ago, an international seminar on 'cancer of the liver' was held in Kampala, Uganda, under the auspices of the International Union against Cancer. Since that meeting (UICC, 1957) experimental and clinical scientists have contributed immensely towards research into the causes and biology of human liver cancer in tropical Africa. Although infectious and inflammatory diseases account for a great percentage of deaths among the Africans, neoplasms undoubtedly account for a relatively large proportion of the total mortality and morbidity (*Williams*, 1975a). Indeed, liver cancer is endemic in many areas of sub-Saharan Africa.

Although various pathological conditions (parasitic infections, malnutrition, viral hepatitis and cirrhosis) have been shown to influence the development of liver cancer (*Hutt*, 1971; *Martini*, 1976), there is no evidence which indicates that these conditions are directly responsible for the development of liver cancer in man. The consensus of opinion among cancer researchers is that most cancers are caused by exposure to environmental factors. There is the view that most of the chemically induced cancers are not caused by only one carcinogen. *Schmahl* (1976) has shown that there is a syncarcinogenic action if low dosages of different agents, which have the same organotropy, were applied simultaneously or one after the other. The relative high incidence of hepatocellular carcinoma in Nigeria could have a multiple-factor aetiology. In order to control the disease, therefore, these factors should be identified and eliminated from the Nigerian environment.

A. Role of Carcinogen Poisoning

A critical assessment of the role of ingestion of fungal carcinogens in the aetiology of human liver cancer was published recently by *Wogan* (1975a). One of the main recommendations made by the Sub-Committee of IARC (1971) was that the actual carcinogenic role, in man, of the total environmental and plant carcinogens should be determined in different populations at risk.

In spite of the recent revelations (*Alpert et al.*, 1971; *Shank et al.*, 1972a, b; *Peers and Linsell*, 1973; *Van Rensburg et al.*, 1974, 1975; *Wogan*, 1975b; *Peers et al.*, 1976) that there is an association between aflatoxin content of food and hepatoma frequency in several countries, there are insufficient data on the involvement of aflatoxin (or other carcinogens) in the development of liver cell carcinoma in a high incidence area such as Nigeria. Research in this area has, so far, not yielded meaningful results because of (a) lack of development of reliable assay techniques and (b) inadequacies in the analytical methods in detection of carcinogens (or their metabolites) in biological tissues. However, the identification of aflatoxin and its metabolite in urine samples of normal individuals and of patients with liver diseases in Ibadan (*Bababunmi et al.*, 1976) using sensitive fluorimetric method (*Bababunmi et al.*, 1975) further confirm that carcinogen-contaminated diets (*Bababunmi*, 1976) are ingested by the Nigerian. Reliable techniques for the detection of nitrosamines are gradually being standardized (*Brooks et al.*, 1972; *Cox*, 1973; *Fiddler*, 1975). The fact that there is undoubtedly nitrosamine contamination of some local beverages coupled with the observation of *Montesano and Magee* (1971, 1974) that there is a good correlation between metabolism of nitrosamines and tumour induction in animal models suggests the possible involvement of nitrosamine as a contributory aetiological factor to the human liver cancer in Nigeria. In rats, liver tumours are the most frequent neoplasms obtained with dimethylnitrosamine (*Shank*, 1975). Adequate epidemiological data are, however, required to support the possible role of nitrosamine poisoning in the high incidence of liver cancer disease in localized regions of Africa (*Doll*, 1968).

Native brews and major herbal preparations are drunk daily by all sections of the Nigerian community. The detection of carcinogenic substances such as retrorsine and monocrotaline in these plants is an indication of their potential role as syncarcinogens in humans.

B. Other Possible Influential Factors

On the whole, the possible environmental agents which are oncogenic in tropical Africa have not been identified. Apart from the possible role of hepatitis B surface antigen in the aetiology of liver cell carcinoma (*Williams*, 1975b), there are other types of potential cancer-inducing agents, such as ionizing radiation, which should be evaluated. The recent review of *Heidelberger* (1975) highlights various mechanisms.

In many countries in tropical Africa, there is a reckless dispensation of drugs and medicines (in such forms as antibiotics) are hawked in the streets, consumed by the common Nigerian without a proper prescription from a qualified physician. It is known that some antibiotics such as daunomycin and adriamycin are carcinogenic in rats (*Bertazzoli et al.,* 1971). Although there is no available test for liver carcinogenicity of antibiotics, there are some case reports (*Mukherji,* 1957; *Cohen and Greger,* 1967; *Fraumeni,* 1967; *Awwaad et al.,* 1975) which suggest that aplastic anaemia caused by chloramphenicol is associated with the subsequent development of myeloblastic leukaemia.

Other carcinogenic metabolites of fungi, such as sterigmatocystin (*Wogan,* 1975a), should be screened for in local foodstuffs. Known toxic metabolites of *Aspergillus* such as palmotoxin (*Bassir and Adekunle,* 1968, 1970; *Bassir and Emerole,* 1974, 1975) should be tested for carcinogenicity. There are many other different toxic substances (*Oke,* 1969) to which the Nigerian is exposed. Some of these are the endogenous toxins which occur naturally in foods; for example, dioscorine which is the alkaloid responsible for the toxicity of some wild yams (*Dioscorea* spp.) of West Africa (*Nicholis et al.,* 1961; *Olaniyi,* 1973). Some edible mushrooms (*Oso,* 1975) in Nigeria are toxic. Relatively large quantities of pepper are used in the preparation of sauce in Nigeria. Species of pepper such as *Piper nigrum* contains alkaloids and volatile oils which are toxic to both animals and man. Some tropical foodstuffs such as soyabean, breadfruit and melon contain some levels of sapotoxin which have toxic effects on man. Indeed, there is a need to estimate the total load of carcinogens in the Nigerian environment.

Apart from foods, beverages and pharmaceuticals, there are other sources of potential carcinogens which are introduced by man into the environment in such forms as cosmetics, toileteries and food additives. With the arrival of various industries in Nigerian cities, inhalation of dust and vapours present a new form of danger. Epidemiological appraisal of these possible factors is lacking and clinical implications remain to be resolved.

VII. Conclusion

The isolation and identification of toxins and potentially carcinogenic substances represent a formidable task. Patients with liver cancer usually cannot be identified prior to the development of the disease. All over the world, researchers are faced with legal restrictions and ethics which limit experimentation with human subjects. Therefore, the only way to identify cancer-inducing substances is by using well-designed and valid microorganism and animal assays. The use of rapid test methods (*Slater et al.,* 1971; *De Serres,* 1974; *Shubik,* 1975) and animal models (*Conney and Levin,* 1974) should be maximized in the screening of environmental substances for the presence of carcinogens. The rapid test methods of *Ames et al.* (1973) are mutagenicity testing in bacterial systems. Detection of carcinogens as mutagens in the bacterium/mammalian microsome has been discussed by *Ames et al.* (1975) and *McCann and Ames* (1976). Although there is a high degree of correlation between results obtained in animal model tests and mutagenicity studies, a carcinogen is not necessarily a mutagen or vice versa (*Shubik*, 1975). The very recent excellent review of *Drake and Blatz* (1976) gives prominence to biochemical aspects of mutagenesis.

The fact that animal species vary in their responses (*Williams,* 1959; *Parke*, 1968; *Smith*, 1973) to toxic substances is a matter of considerable importance when animals are used to evaluate the carcinogenicity of environmental factors to which human beings may be exposed. In all toxicological studies, the dietary habit of the animal model is an important factor (*Tannenbaum and Silverstone*, 1953). How dietary factors can affect the development of primary liver cancer is indeed a question yet to be answered in many tropical countries of Africa.

References

Aboyade, O.: Nigeria's economy; in Africa South of the Sahara; 1st ed., pp. 558–563 (Europa Publications, London 1971).

Adeniji, M.O.: Role of fungi associated with brown root rot of tomato in *Colletotrichum* infection. Nig. agric. J. *3:* 24–31 (1966).

Adeniji, M.O.: Fungi associated with storage decay of yam in Nigeria. Phytopathology *60:* 590–592 (1970).

Adeniji, M.O.: Fungi associated with the deterioration of gari. Nig. J. Plant Protect. (in press, 1976).

Akinrele, I.A.: Fermentation of cassava. J. Sci. Food Agric. *15:* 589–594 (1964).

Alpert, M.E.; Hutt, M.S.R.; Wogan, G.N., and Davidson, C.S.: Association between aflatoxin content of food and hepatoma frequency in Uganda. Cancer *28:* 253–260 (1971).

Ames, B.N.; Durston, W.E.; Yamasaki, E., and Lee, F.D.: Carcinogens are mutagens: a simple test system combining liver homogenates for activation and bacterial for detection. Proc. natn. Acad. Sci. USA *70:* 2281–2285 (1973a).

Ames, B.N.; Lee, F.D., and Durston, W.E.: An improved bacterial test system for the detection and classification of mutagens and carcinogens. Proc. natn. Acad. Sci. USA *70:* 782–786 (1973b).

Ames, B.N.; McCann, J., and Yamasaki, E.: Methods for detecting carcinogens and mutagens with the salmonella/mammalian microsome mutagenicity test. Mutation Res. *31:* 347–363 (1975).

Awwaad, S.; Khalifa, A.S., and Kamel, K.: Vacuolization of leukocytes and bone marrow aplasia due to chloramphenicol toxicity. Clin. Pediat. *14:* 499–506 (1975).

AOAC: Natural poisons; in *Horwitz* Official methods of analysis (Association of Official Analytical Chemists, Washington 1975).

Asplin, F.D. and Carnaghan, R.B.A.: The toxicity of certain groundnut meals for poultry with special reference to their effect on ducklings and chickens. Vet. Rec. *73:* 1215–1219 (1961).

Ayanaba, A. and Alexander, M.: Microbial formation of nitrosamines *in vitro*. Appl. Microbiol. *25:* 862–867 (1973).

Aykroyd, W.R. and Joyce, D.: Legumes in human nutrition, p. 138 (FAO, Roma 1964).

Bababunmi, E.A.: Excretion of aflatoxin in the urine of normal individuals and patients with liver diseases in Ibadan (Nigeria); in *Nieburgs* Detection and prevention of cancer (Decker, New York 1976).

Bababunmi, E.A.; Francis, T.I.; Bassir, O., and Rukari, A.: Urinary excretion of aflatoxin by Nigerians with acute liver diseases; in Proc. 3rd Int. Symp. De PCa, New York 1976, abstr.

Bababunmi, E.A.; French, M.R.; Rutman, R.R.; Bassir, O., and Dring, L.G.: A study of the fluorescence of 5-hydroxycoumarin, 5-methoxycoumarin and aflatoxin B_1. Biochem. Soc. Trans. *3:* 940–943 (1975).

Bassir, O.: Toxic substances in Nigerian foods. West Afr. J. Biol. appl. Chem. *12:* 3–6 (1969).

Bassir, O. and Adekunle, A.A.: Two new metabolites of *Aspergillus flavus* (Link). FEBS Lett. *2:* 23–25 (1968).

Bassir, O. and Adekunle, A.A.: Teratogenic action of aflatoxin B_1 and palmotoxins B_0 and G_0. J. Path. *102:* 49–52 (1970).

Bassir, O. and Emerole, G.O.: Metabolism of palmotoxins B_0 and G_0 *in vitro*. Eur. J. Biochem. *47:* 321–324 (1974).

Bassir, O. and Emerole, G.O.: Species differences in the metabolism of palmotoxins B_0 and G_0 *in vitro*. Xenobiotica *5:* 649–655 (1975).

Bertazzoli, C.; Chieli, T., and Solcia, E.: Different incidence of breast carcinomas or fibroadenomas in daunomycin or adriamycin-treated rats. Experientia *27:* 1209–1210 (1971).

Blount, W.P.: Turkey 'X' disease. Turkeys. J. Br. Turkey Fed. *9:* 52–77 (1961).

Bogovski, P.; Preussmann, R.; Walker, E.A., and Davis, W.: IARC scientific publications No. 3 (International Agency for Research on Cancer, Lyon 1972).

Bras, G.; Berry, D.M., and Gyorgy, P.: Plants as aetiological factor in veno-occlusive disease of the liver. Lancet *i:* 960–962 (1957).

Brooks, J.B.; Alley, C.C., and Jones, R.: Reaction of nitrosamine with fluorinated anhydrides and pyridine to form electron capturing derivatives. Analyt. Chem. *44:* 1881–1884 (1972).

Ciegler, A.; Peterson, R.E.; Lagoda, A.A., and Hall, H.H.: Aflatoxin production and degradation by *Aspergillus flavus* in 20-litre fermentors. Appl. Microbiol. *14:* 826–833 (1966).

Cohen, T. and Greger, W.P.: Acute myeloid leukaemia following seven years of aplastic anaemia induced by chloramphenicol. Am. J. Med. *43:* 762–770 (1967).

Conney, A.H. and Levin, W.: Carcinogen metabolism in experimental animals and man; in IARC scientific publications No. 10, pp. 3–24 (International Agency for Research on Cancer, Lyon 1974).

Coursey, D.G. and Haynes, P.H.: Root crops and their potential as food in the tropics; in World crops, pp. 1–5 (Morgan Grampian, London 1970).

Cox, G.B.: Estimation of volatile *N*-nitrosamines by high-performance liquid chromatography. J. Chromat. *83:* 471–481 (1973).

Dalziel, J.M.: Useful plants of west tropical Africa. Appendix in *Hutchinson and Dalziel* Floral of west tropical Africa (Crown Agents for the Colonies, London 1948).

Dickens, F.: Aflatoxins: a carcinogen present in infected groundnuts. Br. med. Bull. *20:* 96–101 (1964).

Dickens, F. and Jones, H.E.M.: Further studies on the carcinogenic and growth inhibitory activity of lactones and related substances. Br. J. Cancer *17:* 100–108 (1963).

Diener, U.L. and Davis, N.D.: Effect of environment on aflatoxin production in freshly dug peanuts. Trop. Sci. *10:* 22–28 (1968).

Diener, U.L. and Davis, N.D.: Limiting temperature and relative humidity for aflatoxin production by *Aspergillus flavus* in stored peanuts. J. Am. Oil Chem. Soc. *47:* 347–351 (1970).

Doll, R.: Cancer in Africa (East African Publishing House, Nairobi 1968).

Doll, R.; Payne, P.M., and Waterhouse, J.A.H.: Cancer incidence in five continents. UICC publication (Springer, Berlin 1966).

Drake, J.W. and Baltz, R.H.: The biochemistry of mutagenesis. Annu. Rev. Biochem. *45:* 11–37 (1976).

Edington, G.M. and MacLean, C.M.U.: A cancer rate survey in Ibadan. Br. J. Cancer *19:* 471–481 (1965).

Ender, F.; Havre, G.; Helgebostad, A.; Koppang, N.; Madsen, R., and CEH, L.: Isolation and identification of a hepatotoxic factor in herring meal produced from sodium nitrite preserved herring. Naturwissenschaften *24:* 637–638 (1964).

Fafunso, M. and Bassir, O.: Nutritional qualities of some African leafy vegetables. Effect of boiling on the essential amino acid composition of their extracted protein. J. Food Sci. *41:* 214–215 (1976).

Fiddler, W.: The occurrence and determination of *N*-nitroso compounds. Toxic. appl. Pharmac. *31:* 352–360 (1975).

Fraumeni, J.F.: Bone marrow depression induced by chloramphenicol or phenylbutazone. J. Am. med. Ass. *201:* 150–156 (1967).

Heidelberger, C.: Chemical carcinogenesis. Annu. Rev. Biochem. *44:* 79–121 (1975).

Hodges, F.A.; Zust, J.R.; Smith, H.R.; Nelson, A.A.; Armbrecht, B.H., and Campbell, A.D.: Mycotoxins: aflatoxin isolated from *Penicillium puberulum*. Science *145:* 1439 (1964).

Hutt, M.S.R.: Epidemiology of human primary liver cancer; in IARC scientific publications No. 1, pp. 21–29 (International Agency for Research on Cancer, Lyon 1971).

IARC: Liver cancer. Scientific publications No. 1, pp. 171–176 (International Agency for Research on Cancer, Lyon 1971).

IARC: Evaluation of carcinogenic risk of chemicals to man, vol. 1, pp. 95–106 (International Agency for Research on Cancer, Lyon 1972).

IARC: Evaluation of carcinogenic risk of chemicals to man, vol. 10, p. 305 (International Agency for Research on Cancer, Lyon 1976).

Idusogie, E.O.: Centuries of changing food consumption patterns in African communities; in Food and nutrition in Africa. Bull. FAO/WHO/OAU 12, pp. 5–32 (1973).

Issenberg, P.: Nitrite, nitrosamines and cancer. Fed. Proc. Fed. Am. Socs exp. Biol. *35:* 1322–1326 (1976).

Joaquim, K.: Nitrosamine contamination of some Nigerian beverages; PhD thesis, Ibadan (1973).

Keen, P. and Martin, P.: Is aflatoxin carcinogenic in man? The evidence in Swaziland. Trop. geogr. Med. *23:* 44–53 (1971).

Legator, M.S.: Biological assay for aflatoxins; in *Goldblatt* Aflatoxin, pp. 107–149 (Academic Press, New York 1969).

Lijinsky, W. and Epstein, S.S.: Nitrosamines as environmental carcinogens. Nature, Lond. *225:* 21–23 (1970).

Maduagwu, E.: Nitrosamine contamination of some Nigerian beverages; PhD thesis, Ibadan (1976).

Magee, P.N.: Toxicity of nitrosamines: their possible health hazards. Food Cosmet. Toxicol. *9:* 207–218 (1971).

Magee, P.N. and Barnes, J.M.: Carcinogenic nitroso compounds. Adv. Cancer Res. *10:* 163–246 (1967).

Madhavan, T.V.; Tulpule, O.G., and Gopalan, C.: Aflatoxin-induced hepatic fibrosis in rhesus monkeys. Archs Path. *79:* 466–469 (1965).

Martinin, G.A.: The role of alcohol in the etiology of cancer of the liver. Proc. 3rd Int. Symp. DePCa, New York 1976, abstr.

McCann, J. and Ames, B.N.: Detection of carcinogens as mutagens in the Salmonella/microsome test. Assay of 300 chemicals: Discussion. Proc. natn. Acad. Sci. USA *73:* 950–954 (1976).

McCann, J.; Spingarn, N.E.; Kobori, J., and Ames, B.N.: The detection of carcinogens as mutagens: bacterial tester strains with R-factor plasmids. Proc. natn. Acad. Sci. USA *72:* 979–983 (1975).

Miller, J.A. and Miller, E.C.: Carcinogens occurring naturally in foods. Fed. Proc. *35:* 1316–1321 (1976).

Mirvish, S.S.: Formation of *N*-nitroso compounds: chemistry, kinetics and *in vivo* occurrence. Toxic. appl. Pharmac. *31:* 325–351 (1975).

Montesano, R. and Magee, P.N.: Evidence of formation of *N*-methyl-*N*-nitrosourea in rats given *N*-methylurea and sodium nitrite. Int. J. Cancer *7:* 249–255 (1971).

Montesano, R. and Magee, P.N.: Comparative metabolism *in vitro* of nitrosamines in various animal species including man; in IARC scientific publication No. 10, pp. 39–56 (International Agency for Research on Cancer, Lyon 1974).

Mukherij, P.S.: Acute myeloblastic leukaemia following chloramphenicol treatment. Br. med. J. *ii:* 1286–1287 (1957).

Nicholis, L.; Singlair, H.M., and Jelliffe, D.B.: Food poisoning: wild yam poisoning; in Tropical nutrition and dietetics, pp. 377 (Baillière, Tindal & Cox, London 1961).

Odebiyi, A.I.: Demographic and socio-economic aspects of cancer in the city of Ibadan; MSci thesis, Ibadan (1972).

Oke, O.L.: Oxalic acids in plants and in nutrition. Wld Rev. Nutr. Diet., vol. 8, p. 25 (Karger, Basel 1967).

Oke, O.L.: The present status of nutrition in Nigeria. Wld Rev. Nutr. Diet., vol. 10, pp. 262–303 (Karger, Basel 1969).

Oke, O.L. and Umoh, I.B.: Nutritive value of leaf protein: a note on the comparison of *in vitro* and *in vivo* methods. Nutr. Rep. int. *10:* 397–403 (1974).

Okorie, D.A.: Atanisatin: a carbazole alkaloid. Nig. J. Sci. *9:* 201–210 (1975).

Olaniyi, O.A.: Studies on the browning of yam flour; PhD thesis, Ibadan (1973).

Olayide, S.O.; Olatunbosun, D.; Idusogie, E.O., and Abiagom, J.D.: A quantitative analysis of food requirements, supplies and demands in Nigeria, 1968–1985 (Publ. Fed. Dept. Agric., Lagos 1972).

Oso, B.A.: Mushrooms and the Yoruba people of Nigeria. Mycologia *67:* 311–319 (1975).

Oyeniran, J.O. and Adeniji, M.O.: Mould deterioration of cocoa during processing and storage in Nigeria. Nig. J. Plant Protect. *1:* 48–51 (1975)

Oyenuga, V.A.: Nigeria's foods and feeding-stuffs (Ibadan University Press, Ibadan 1968).

Parke, D.V.: The biochemistry of foreign compounds (Pergamon Press, Oxford 1968).

Peers, F.G.; Gilman, G.A., and Linsell, C.A.: Dietary aflatoxins and human liver cancer. A study in Swaziland. Int. J. Cancer *17:* 167–176 (1976).

Peers, F.G. and Linsell, C.A.: Dietary aflatoxins and liver cancer. A population based study in Kenya. Br. J. Cancer *27:* 473–484 (1973).

Purchase, I.F.H.: Cancer in Africa (East African Publishing House, Nairobi 1968).

Purchase, I.F.H. and Watt, J.J. van der: The aetiology of primary liver cancer in the Bantu. J. Path. *102:* 163–169 (1970).

Rensberg, S.J. van; Watt, J.J. van der; Purchase, I.F.H.; Coutinho, L.P., and Markham, R.: Primary liver cancer rate and aflatoxin intake in a high cancer area. S. Afr. med. J. *48:* 2508a–d (1974).

Rensberg, S.J. van; Kirsipuu, A.; Coutinho, L.P., and Watt, J.J. van der: Circumstances associated with the contamination of food by aflatoxin in a high primary liver cancer area. S. Afr. med. J. *49:* 877–883 (1975).

Rohrborn, G.: Mutagenesis and carcinogenesis; in IARC scientific publications No. 10, pp. 213–219 (International Agency for Research on Cancer, Lyon 1974).

Schmahl, D.: Multiple factor etiology: an overview; in Proc. 3rd Int. Symp. DePCa, New York 1976, abstr.

Schoental, R.: Induction of tumours and bifunctional crosslinking metabolites of nitrosamines. Br. J. Cancer *33:* 668–669 (1976).

Semeniuk, G.: Microflora; in *Anderson and Alcock* Storage of cereal grains and their products, pp. 77–151 (Am. Ass. Cereal Chemists, St. Paul 1954).

Serres, F.J. de: Mutagenic specificity of chemical carcinogens in micro-organisms; in IARC scientific publications No. 10, pp. 201–211 (International Agency for Research on Cancer, Lyon 1974).

Shank, R.C.: Toxicology of *N*-nitroso compounds. Toxic. appl. Pharmac. *31:* 361–368 (1975).

Shank, R.C.; Bhamarapravati, N.; Gordon, J.E., and Wogan, G.N.: Dietary aflatoxins and human liver cancer. IV. Incidence of primary liver cancer in two municipal populations of Thailand. Food Cosmet. Toxicol. *10:* 171–179 (1972).

Shank, R.C.; Gibson, J.B.; Nondasuta, A., and Wogan, G.N.: Dietary aflatoxins and human liver cancer. II. Aflatoxins in market foods and foodstuffs of Thailand and Hong Kong. Food Cosmet. Toxicol. *10:* 61–69 (1972a).

Shank, R.C.; Gordon, J.E.; Nondasuda, A.; Subhamani, B., and Wogan, G.N.: Dietary aflatoxins and human liver cancer. III. Field survey of rural Thai families for ingested aflatoxins. Food Cosmet. Toxicol. *10:* 71–84 (1972b).

Shubik, P.: Potiential carcinogenicity of food additives and contaminants. Cancer Res. *35:* 3475–3480 (1975).

Slater, E.E.; Anderson, M.D., and Rosenkranz, H.S.: Rapid detection of mutagens and carcinogens. Cancer Res. *31:* 970–973 (1971).

Smith, R.L.: The excretory function of bile (Chapman & Hull, London 1973).

Tannenbaum, A. and Silverstone, H.: Nutrition in relation to cancer. Adv. Cancer Res. *1:* 451–501 (1953).

UICC (International Union against Cancer): Acta Un. Int. Cancer *13:* 516 (1957).

Williams, A.O.: Tumours of childhood in Ibadan, Nigeria. Cancer *36:* 370–378 (1975a).

Williams, A.O.: Hepatitis B surface antigen and liver cell carcinoma. Am. J. med. Sci. *270:* 53–56 (1975b).

Williams, A.O.; Edington, G.M., and Obakponovwe, O.: Hepatocellular carcinoma in infancy and childhood. Br. J. Cancer *21:* 474–482 (1967).

Williams, R.T.: Detoxication mechanisms; 2nd ed. (Chapman & Hull, London 1959).

Wogan, G.N.: Chemical nature and biological effects of the aflatoxins. Bact. Rev. *30:* 460–470 (1966).

Wogan, G.N.: Mycotoxins. Annu. Rev. Pharmac. *15:* 437–451 (1975a).

Wogan, G.N.: Dietary factors and special epidemiological situations of liver cancer in Thailand and Africa. Cancer Res. *35:* 3499–3502 (1975b).

Wolff, I.A and Wasserman, A.E.: Nitrates, nitrites and nitrosamines. Science *117:* 15–19 (1972).

Wynder, E.L.: Introductory remarks: conference of nutrition in the causation of cancer. Cancer Res. *35:* 3238–3239 (1975).

Dr. *E.A. Bababunmi,* Department of Biochemistry, University of Ibadan, *Ibadan, Oyo State* (Nigeria)

Wld Rev. Nutr. Diet., vol. 28, pp. 210–235 (Karger, Basel 1978)

Carcass Evaluation of Cattle, Sheep and Pigs

Alastair Cuthbertson

Meat and Livestock Commission, Bletchley, Milton Keynes

Contents

I. Introduction

It is appropriate at the outset to consider the meaning and relevance of carcass evaluation. Although a literal interpretation of the term implies the process of establishing the monetary value of carcasses, in practice the expression usually refers to the description of the physical characteristics of carcasses. Quite clearly, however, such a description has an important bearing on the financial value placed on carcasses in commercial trading.

Carcasses are evaluated for a variety of purposes and the degree of detail used to describe the physical characteristics depends on the needs of each situation. At all stages in the marketing chain, from farm through to consumer, carcass evaluation has an important part to play. It is of importance to the retailer who has to meet his customer requirements in terms of such characteristics as the size and leanness of joints, and who has to estimate how much meat he will be able to sell from a given carcass. It is, therefore, also of relevance to the wholesaler or packer who buys from farmers and has to meet retailers' needs and, not least, to the farmer who should be concerned to gear his breeding and production system to satisfy meat trade requirements.

Apart from its role in helping to value carcasses in day-to-day trading, carcass evaluation has other important related uses. These are primarily concerned with determining the genetic merit of breeding stock so far as carcass characteristics are concerned, and with the efficiency of lean meat production. This involves selection within breeds on the basis of performance and progeny test results, and selection between breeds using results from breed comparison trials. It implies the need to assess carcass merit in the live animal in certain situations. Such carcass evaluation work can extend to studies of the relative efficiencies of different animals during various stages of growth and how these are affected by nutrition and other environmental factors.

The distinction between carcass and eating quality characteristics is difficult to define because they are closely linked but, as far as possible, this paper will deal with the variables important in preparing meat up to the point of retail sale and omit consideration of the effect of these and other characteristics on eating quality.

Other authors have reviewed for either cattle, sheep or pigs the subject dealt with here (*Carroll and Conniffe*, 1968; *Timon*, 1968; *Osinska*, 1968; *Preston and Willis*, 1974; *Berg and Butterfield*, 1976). Such reviews have brought together, and interpreted as a whole, much of the valuable published information on the growth and development of farm animals and on how differences in carcass composition may be predicted. In preparing this review covering the three species, the author has drawn quite extensively on data from GB's Meat and Livestock Commission (MLC). It seemed appropriate to use these data to illustrate some of the topics which will be discussed since the MLC has been building up very comprehensive carcass data on large numbers of cattle, sheep and pigs in recent years, and the results are only just beginning to be published.

II. Carcass Characteristics and Their Significance

The characteristics which are generally considered to be important in evaluating a carcass, regardless of species, are: (a) weight; (b) percentage of tissues in the carcass, especially of lean meat or fat; (c) distribution of tissues through the carcass, and (d) thickness of tissues. Before starting an examination of alternative evaluation procedures, the relevance of these characteristics will be considered.

A. Weight
The weight or size of a carcass clearly has a major influence not only on the quantity of the various tissues, but also on the size of the muscles exposed on cutting, and of the joints. This is of importance, for example, in relation to a retailer's ability to use his usual cutting procedures on carcasses outside his normal weight specification. However, it is possible to operate cutting techniques to produce acceptable joint sizes for the consumer from carcasses of widely different weight, but the retailer has to be prepared to experiment and the housewife prepared to buy joints which may differ in their form of presentation from previous purchases.

The weight at which an animal is slaughtered depends largely on the weight achieved when it reaches a given level of fatness. This level will depend on the demands of the market and the price differentials associated with different fat levels. Breed, sex and diet/feeding system have a marked effect on the weight achieved.

The effect of breed is very important and the stimulus to innovate in meat cutting and preparation may arise if there is an increasing use of larger, later maturing (or fattening) breeds of beef and sheep. Studies are now in progress in a number of countries to assess the efficiency of different breeds, not only in terms of their growth rate and food conversion efficiency, but also in terms of their ability to turn food into meat. It remains to be seen whether the larger beef breeds, e.g. Charolais, turn out to provide a better overall use of resources than the smaller, earlier maturing breeds such as Aberdeen Angus which would tend to provide a quicker return on investment. While there may be some increase in carcass weight due to breed, there is a trend in some countries to slaughter animals at lower levels of fatness. Due to the positive correlation between weight and fatness, at least within breeds, this seems likely, in the short term, to oppose much overall change in carcass weight in these countries.

An illustration of the differences between some beef breeds commonly used

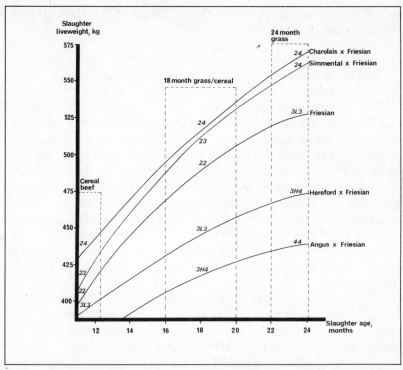

Fig. 1. Average live weight and age at slaughter for steers of several breeds and crosses on different production systems. Typical carcass classification for fatness and conformation is indicated in italics.

in Britain is contained in figure 1. This is based on data obtained from commercial beef units operating intensive (cereal), semi-intensive (grass/cereal) or 2-year systems of production and recording with the MLC. This figure illustrates differences in live weight at slaughter which are associated with similar differences in carcass weight. Also indicated is evidence on the average fatness and conformation of the cattle at slaughter described in terms of the beef carcass classification scheme operating in Britain. The scheme describes carcasses by their weight, fatness, conformation, sex and age group. Fatness is assessed on a scale of five basic classes, where 1 = very lean and 5 = very fat; the central fat class (3) is divided into L (low) and H (high). Carcass conformation or shape (i.e. thickness of lean meat plus fat relative to bone size) is also assessed on a scale of five main classes, where 1 = poor and 5 = very good. The fat class is always referred to before the conformation class so, for example, a carcass described as 3L3 falls in fat class 3L and conformation class 3.

So far as pigs are concerned, the demise of the earlier maturing pig breeds, at least in Britain, has led to pigs with less variation in mature size compared with the ranges for beef and sheep.

B. Tissue Percentages

Within carcasses of similar weight, the percentage formed by the tissues varies considerably and this variation is important. This does not always seem to be recognised by those who point, for example, to high correlations between weight and fatness derived from animals varying widely in weight. It is well established that as the individual animal grows and increases in weight, the proportion of fat increases primarily at the expense of lean meat, but this does not mean that at any one weight all animals of a particular species have the same composition. The results reported by *Kempster et al.* (1976a) provide an illustration of the variation in leanness which is still left amongst lamb carcasses of similar weight. In their survey of the composition of commercial lambs of widely different breed types, the standard deviation of lean content over all lamb groups was 3.96 percentage units. After adjusting the data to constant carcass weight, the standard deviation of lean content was reduced to only 3.78 percentage units.

The percentage of lean meat or fat in carcasses of similar weight is of particular significance since studies in Britain have shown that the leanness of meat is the major criterion by which consumers judge quality over the shop counter (*Brayshaw et al.,* 1965). While there is still a good deal of discussion about the level of fatness required in meat for satisfactory eating quality, and some lay great emphasis on the importance of visible intramuscular or marbling fat in beef, there is growing evidence which fails to support the belief that fatness contributes to the eating quality of lean meat (*Rhodes,* 1976). In general, it appears that the trend towards the production of leaner carcasses is unlikely to threaten the acceptability of the product, except of course to those consumers whose preference is for fat. The revised standards for the United States Department of Agriculture's (USDA) beef grading scheme introduced in 1976 go some way to acknowledging that high fatness is unnecessary for satisfactory eating quality.

The movement towards the production of leaner carcasses, which is evident at least in Britain, will accelerate if fears about the link between saturated animal fats and blood circulatory problems in humans are confirmed, and as producers and the meat industry become aware of the wastefulness of producing fat on animals beyond an optimum. Excess fat produced is normally trimmed and de-valued at the abattoir, retail shop or dinner plate before consumption. To illustrate the differences between the fatness of carcasses from one country to another, the typical commercial beef carcass in Germany, for example, probably contains about 15–20% of physically separable fat compared with some 25% for Britain and over 30% for the USA.

Table I. Approximate composition of beef, sheep and pig carcasses[1] of average fatness in Britain

	Beef	Sheep	Pig
Lean meat, %	59	55	54
Total fat, %	25	28	35
SF	9	13	25
IF	12	$11^1/_2$	7
Perinephric and retro-peritoneal	4	$3^1/_2$	3
Bone (including small waste component), %	16	17	11
Total	100	100	100

[1] All carcasses ex head, feet and skin.

In order to assess for beef the extent of waste fat production in Britain, MLC has recently calculated that if the average consumer requires a 10:1 ratio of lean to fat in meat, then some 230,000 t of fat (both dissectable carcass fat and internal body fat) in excess of requirements was produced in 1975 (*Harrington and Kempster,* 1977). This figure includes fat which may have found a residual use. If fat class 2 in the carcass classification scheme referred to above (which contains 16–20% dissectable carcass fat), is taken as the current level of fatness at optimum efficiency, then 25% of the excess fat produced in 1975 was estimated to be true waste.

The relative contribution of each tissue to the whole carcass by species is illustrated in table I. This shows the way in which a typical beef, sheep and pig carcass in Britain breaks down into its major physically separable components. A pig carcass normally includes the head, feet and skin whereas these are normally excluded from beef and sheep carcasses, and so to allow comparisons between species these components have been omitted from each. It is interesting to observe that pig carcasses contain the highest percentage of edible components (lean + fat) and that beef carcasses contain on average the most lean and least fat. Pig carcasses have a markedly lower bone content than either beef or lamb, but this appears to be at the expense of over a third of the carcass weight being fat.

The values in table I relate to the total percentage of lean meat, fat and bone in the carcass and not to the amalgam of these which is normally bought by consumers. This approach to carcass composition has been adopted because market requirements vary enormously within as well as between countries, and the percentage of the carcass sold to consumers depends on the extent to which they are prepared to accept fat in joints and on the extent to which bone is normally sold in the joints. To illustrate the effect of different procedures, some

70% of the typical beef carcass in Britain is sold as 'saleable meat' across the counter. In this, there is a ratio of lean to fat of approximately 6:1 and little, if any, bone. In contrast, lamb which is traditionally sold with all the bone produces some 90% of 'saleable meat'.

There is considerable variation about the average values shown in table I. For example, in commercial pig carcasses in Britain, where a significant increase in lean content has occurred over the years, some two thirds of carcasses have lean percentages within ± 5 percentage units of the mean, while a comparable figure for cattle and sheep is ± 8 percentage units. These differences in lean content would result in similar variability in saleable meat content on the assumption that the saleable meat from carcasses of different fatness were sold with the same ratio of lean to fat. In practice this does not happen since retailers tend to sell the meat from fatter carcasses with a lower ratio of lean to fat.

C. Tissue Distribution

This is a subject which has evoked much debate over the years between scientists and those commercially involved in the production and marketing of meat. Discussion has been active because, contrary to growing scientific evidence, many in the meat industry still attach great importance to the percentage of higher valued cuts and the distribution of meat through the carcass. Special attention is paid to conformation or shape which is considered to provide a means of assessing variations in these characteristics. Breeders too have focussed attention on conformation as livestock show standards illustrate. This has been done not only because they believe there may be some link with liveability but because of the above meat trade beliefs about its value.

Detailed carcass studies carried out in a number of countries on widely different breeds within species shows, for example, that while there is variation in the distribution of muscles between one part of the carcass and another, the extent of the variation is low and much less than has often been supposed to be the case (*Butterfield*, 1963; *Richmond and Berg*, 1971). Results of comparisons between species have also been reported by *Berg and Butterfield* (1976).

The MLC has recently been exploring differences in the distribution of lean and fat, particularly in beef. In a study covering a wide range of beef carcass types, *Kempster et al.* (1976d) found that when the weight of lean was adjusted to the same level, the difference in the weight of lean meat in the higher priced joints between extreme breed types (Limousin × Friesian and Ayrshire) was 2 kg in the side. This represents a difference of about 1.5% in the retail realisation value of the carcass. It was also found that while variability in the distribution of fat within depots was low, analyses based on a similar sample of carcasses indicated important differences in the way in which fat is partitioned between depots (*Kempster et al.*, 1976b, c). For example, at the same total fat content, carcasses from Ayrshire cattle in the sample contained less subcutaneous fat (SF) and more intermuscular fat (IF), perinephric and retro-peritoneal fat than those from Friesian or beef breed crosses. Such differences are of importance when predicting total fat or intermuscular fat content from SF development. *McClelland and Russel* (1972) have reported significant differences in fat distribution between depots when two breeds of sheep were compared at the same level of total fatness.

D. Thickness of Tissues

Marked variations exist in the thickness of lean meat or of flesh (i.e. lean meat plus fat) on the exposed cut surfaces of joints. Provided it can be achieved without having to trim off excess fat, retailers tend to favour carcasses with good meat thickness because it is considered to improve the appearance of joints. In certain situations, however, the cross-section of *m. longissimus dorsi (m.l.d.)* may be too large and thick (e.g. in pigs over 100 kg carcass weight). In the home, there may be some advantage in thick flesh, such as reduced cooking losses and easier carving, but there is little, if any, evidence to support this. Indeed, the study by *Boccard and Radomska* (1963) which examined the losses in cooking legs of lamb from 3 breeds of markedly different conformation failed to demonstrate a significant difference between the types represented.

Meat traders in many countries traditionally attach considerable importance to the role of conformation in indicating carcasses with good meat thickness amongst those of similar weight. The extent to which conformation is of value in distinguishing between carcasses for various characteristics, including meat thickness, is being assessed by MLC. So far, studies on beef and lamb have indicated that a judgement of overall conformation explains less than 20% of the variation in area of *m.l.d.* after eliminating the effects of weight and SF content. As might be expected, there is a trend for carcasses of good conformation to have larger muscle cross-sections and to be heavier per unit length, but the correlations are not sufficiently good to provide a reliable guide to muscle thickness on individual carcasses.

It seems clear that the economic significance of variations in meat thickness is much less important to the meat industry than, for example, variation in fat content which has an immediate impact on realisation values.

III. Evaluation Techniques

A. General

From the discussion so far, it will be evident that studies involving carcass evaluation are best concentrated on those characteristics which have the greatest effect on carcass realisation value. Setting aside weight, the description of carcass leanness or fatness is of first importance, and much of the remainder of this chapter will be concerned with techniques for assessing the lean content of carcasses under different conditions.

As a survey of the literature will show, there is a great deal of published work in which a wide range of methods have been applied to assess variations in carcass leanness or, as is common in North America, trimmed, boneless cuts. The method used will depend on the accuracy and precision of the answers required and on the resources available to cover labour, equipment and carcass depreciation costs.

B. Establishing the Base-Line

Most techniques use as a base-line the lean content of the carcass obtained by physical separation after preliminary jointing, but the chemical composition of the carcass is sometimes used, particularly in nutritional studies where it may be important to study the extent to which specific nutrients are retained in the carcass and in the different tissues. For most purposes, the base-line of physical separation seems preferable since carcass value judgements by consumers are made on the basis of the physical appearance or composition of the carcass or joint. In addition, chemical analysis does not appear to be cheaper to operate as a base-line if sampling procedures such as that described by *Florence and Mitchell* (1972) were adopted.

If resources allow, the most comprehensive procedure is to apply a physical separation technique after cutting the carcass into joints followed by chemical analysis of the resulting tissues. Such an approach to providing the base-line would be particularly valuable in growth and nutritional studies and was one of the recommendations made by an EEC group of experts considering carcass evaluation procedures for beef (Commission of the European Communities, 1976). Apart from providing information on the chemical composition of the tissues, such a procedure also enables the standard of physical separation to be monitored.

The physical separation techniques used to determine the base-line include careful separation, with scalpels and scissors, of individual muscles from bone attachments and surrounding fat (*Pomeroy et al.,* 1974) and methods involving separation by butcher's knife into lean meat, fat and bone within major commercial joints (*Cuthbertson et al.,* 1972). Less detailed techniques widely used in the USA, particularly for beef and pork, involve the separation of carcasses in a commercial, but to a degree standardised, manner by jointing and then deboning and trimming excess fat. A technique sometimes used is to separate the soft tissues from the bones followed by a rough (or detailed) separation of the muscle and fat tissues. The fat content of the muscle mass is then determined chemically allowing the fat-free muscle mass to be calculated. The range of separation procedures used for beef by research groups in Europe has been summarised by the EAAP (1976).

The separation technique described by *Cuthbertson et al.* (1972) has been used to provide the base-line for many studies of, for example, differences in carcass composition within and between breeds. It involves weighing the lean meat (including intramuscular fat), the various physically separable fat depots, bone and trimmings within each of a number of standardised commercial joints. By adding together results for the various joints, the composition of the whole carcass is obtained. Cutting sides into joints before subsequent separation of the tissues enables information to be obtained on variations in the proportion of joints from animals of the same and of different breeds and on variations in the distribution and thickness of the tissues in these joints. This technique seems a suitable one to provide a base-line for most studies and it can be linked with chemical analysis where more comprehensive information is required. For those interested in the grouping of muscles anatomically rather than by commercial joints, it is feasible to obtain information on both commercial and anatomical bases by adding together the weight of muscle portions which occur in different commercial joints. In view of the relative constancy of muscle distribution, it seems difficult, except in a few specialised cases, to justify the implementation of more detailed separation procedures especially where it involves the separation and weighing of all the individual muscles and not simply muscle blocks.

Even with the MLC procedure discussed above, considerable costs are involved. At prices in late 1976, the depreciation in carcass value of a side of beef, lamb and pork amounted to £ 26, £ 6 and £ 5, respectively. In addition, there is a heavy labour component which, including preliminary photography and measurement, amounts to some 2.5 man-days for a side of beef and 0.5 and 0.6 man-days for a side of lamb and pork, respectively.

So far as studies involving chemical composition are concerned, the main requirement is to assess the fat, protein, moisture and ash content of the tissues. Standard techniques are available for estimating the above components (AOAC, 1975). The results obtained depend very much on the procedure adopted, and so it is important to quote the technique used when any results are presented. Before such analyses can be carried out, it is vital that the tissues are adequately sampled and that the material analysed is homogeneous.

The procedures adopted to provide the base-line will inevitably be a compromise between the desire to have precise information on individual animals and the need to apply the technique to as many animals as possible to minimise sampling errors.

C. Predicting Base-Line Values

Once a satisfactory procedure has been established for providing the base-line for assessing the important carcass components, the cost of implementing it highlights the need to find methods of predicting overall values by less expensive techniques.

There are several broad approaches ranging from the use of commercial cutting procedures and the separation of sample joints to simple measurements and subjective judgements.

1. Commercial trimming. A commercial type of de-boning and fat trimming procedure, like those used in cutting plants for producing saleable meat for retail sale, is one possible approach. Although such procedures commonly applied in the USA on beef and pork are difficult to standardise and, to some extent, the variations found reflect variations in residual fat in the saleable meat, MLC studies applying such a technique to beef carcasses show that some 80% of the variation in the percentage of lean in side can be explained by the percentage of saleable meat or, indeed, fat trim. An advantage in expressing the results of test work in terms of such characteristics as the quantity of saleable meat is that it can help in communicating the findings to industry. Another advantage of the technique is that it enables joints at the end of the procedure to be sold commercially with little, if any, depreciation.

In some studies, it is important to separate the resulting saleable meat from a sample of the carcasses into lean meat and fat, together with any bone which may be left in the saleable meat, in order to assess the degree of residual fat, and to estimate the regression relationship between, for example, percentage fat trim and percentage total fat. Such detailed work on a sample of carcasses may be justified in, say, breed comparisons where it is desirable to ensure that the commercial cutting procedure is detecting all the differences that are of significance.

2. Sample joints. An alternative procedure is to make use of the close relationship which has been found to exist between the composition of some individual joints and overall carcass composition. Although there have been many studies to compare the value of different joints, such as *Timon and Bichard* (1965a) and *Kempster et al.* (1976a) for lamb, and *Cook et al.* (1974) for pigs, the information upon which to base the selection of a sample joint for a particular application is sometimes limited and confusing.

Recently, *Kempster and Jones* (1977) have examined two important aspects of the problem. These are, firstly, how should the composition of the sample joint be used to give the most precise prediction of carcass lean content? For example, should the prediction equation be constructed with the weight of lean in the joint as an independent variate or with weight of lean in the joint expressed as a percentage of the joint weight as the independent variate? Secondly, to what extent can equations obtained for one sample of carcasses predict the lean content of another sample differing, for example, in breed, sex or plane of nutrition?

Kempster and Jones' (1977) study on different groups of beef cattle indicates that the relative precision of the joints depends on whether or not the prediction equation involves joint weight. When side weight and weight of lean

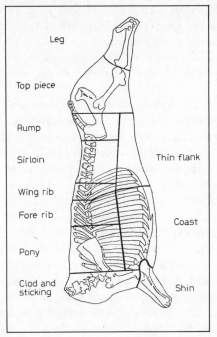

Fig. 2. Location of standardised joints in carcass. From *Kempster et al.* (1976b).

in joints were included as independent variates, there was little difference
between joints in precision, and the smaller joints (shin and leg) appeared best
because they are easy to remove from the carcass and easy to separate — see
figure 2 for location of joints in carcass. The addition of joint weight to the
equation improved considerably the precision of all joints except the shin and
leg. The largest joints, top piece and coast, gave the greatest precision, but the
fore rib probably offered the best compromise between cost and precision. The
within group residual standard deviations (kg lean in side) for the three joints on
this basis were 1.37, 1.34 and 1.58, respectively.

It is important to consider the stability of prediction equations between
groups since, if good stability can be demonstrated for some joints it strengthens
the value of applying a sample joint separation technique to different groups of
animals. *Kempster and Jones* (1977) found that the joints giving the most precise estimate of lean content also tended to be the most stable.

Some support for these findings on beef is provided by the study on lamb
by *Kempster et al.* (1976a).

In deciding which joint to use, it is important to obtain estimates for the
cost of removing the joints under consideration as these can be quite large if

meat traders are unwilling to be left with a carcass minus one joint. In this respect, it is unfortunate that in beef the joints at the extremities of the carcass are not better representatives of overall composition as they would cause the least disturbance. However, in pigs *Cook et al.* (1974) found the hand joint a useful predictor of leanness when combined with fat depth measurements in the loin.

3. Simple measurements and subjective assessments. In many situations, any form of carcass separation is precluded due to cost or practical feasibility. For example, it is not practicable in commercial carcass description or classification schemes to carry out any form of separation other than, perhaps, cutting the loin of one side to examine the *m.l.d.* and the fat over it such as in the Canadian Beef Grading Scheme. In such schemes, it is necessary to strike a balance between a system which provides an objective and accurate estimate of the important features of carcasses, and one which can be applied simply and cheaply to large numbers of carcasses under fast moving slaughter line conditions. This forces one into simple objective measurements and subjective scores, but it is important to make the best choice. The relationship between cost and precision for a number of different predictors of the lean content of beef carcasses is illustrated in figure 3. The results, based on MLC evidence show, for example, that even a subjective judgement of fatness (SF score) by a trained person is better than separating the shin joint. Nevertheless, other things being equal, there are advantages in making the estimation of leanness as objective as possible, but it is important to recognise that many so-called objective measurements have some subjectivity associated with taking them. For example, in measuring the thickness of fat over the *m.l.d.* of intact pig carcasses or sides using an optical probe or intrascope, the actual measurement obtained is influenced by the care taken in inserting the instrument at the correct position and by the pressure applied in taking it. However, it is worth noting that the relationship between, say, a probe measurement on an intact carcass and the corresponding measurement on the cut surface of the carcass may not be relevant in many instances. Instead, it is the relationship between the probe measurement and the percentage of lean in carcass which is important.

The easiest way to predict the percentage of lean meat in a carcass is through the assessment of fatness, especially of SF. The relationships obtained for pigs are better than those for beef and sheep because, firstly, as seen from table I, SF contributes a higher proportion of total fat than in beef and sheep and, secondly, because the SF is spread more thinly and variably over the carcass of beef and lamb. Hence, a simple fat measurement in beef and lamb is likely to give a poorer guide than in pigs, and the value for both beef and lamb can be greatly influenced by the skill applied in removing the skin.

4. Density. The prediction of overall carcass composition using density or specific gravity for cattle, sheep and pigs has been reported by several workers

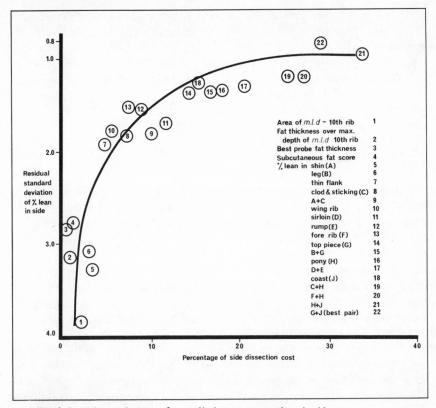

Fig. 3. Precision against cost for predicting percentage lean in side.

(*Adam and Smith,* 1964; *Timon and Bichard,* 1965b). The approach relies on the difference in density between the fat and fat-free parts of the body, in this case assuming a constant composition and density for the two parts. Inaccuracies of the method arise partly from the fact that this is not a completely valid assumption, especially in immature animals. However, the method has been used most often on animals at commercial slaughter weights and, therefore, fairly mature.

The evidence from the various studies carried out is that density is likely to be little better than good fat depth measurements, and MLC evidence using the technique on pigs showed it to be no better than a simple fat measurement over the *m.l.d.* at the head of the last rib and, when combined, added little in predictive value. However, since it provides an index of overall carcass fatness it could be of value, as *Harrington* (1965) has indicated, in a selection programme based on fat depth measurements or sample joint dissection of a rather small portion of the carcass. Without some such validation procedure, the animals with exceptional tissue distribution (e.g. low SF:IF ratio) might be selected over the generations. The progress achieved in carcass leanness in these circumstances would be less than that expected on the basis of the change in the index.

The development of a satisfactory air or gaseous displacement technique seems desirable if the technique is found to be of value and worth using more widely.

IV. Grading and Classification

It will be evident from earlier sections that there is considerable variation in the type of carcass produced in different countries and, even, within the same country. This partly reflects variation in the demand for meat of different qualities and, over the years, has led to schemes being developed by governments, their agencies or the meat trade for identifying different carcass types. These schemes have been mainly carcass grading schemes which group together carcasses with similar characteristics which are considered of most importance to those trading in them, and to which price differentials are then applied. Among the characteristics incorporated in forming the grades, which may number two or more, are weight and fatness.

A problem associated with those grading systems drawn up by individual meat wholesalers to suit their own particular requirements when purchasing from producers, and which should reflect their customers' requirements, is that while each wholesaler may know the type of carcass which should fall into a particular grade, it is difficult for this to be communicated clearly to others. It makes it particularly difficult for the producer who needs to know, for example, the type of carcass he should be producing and to assess where he can sell the types of animals he produces to best advantage. It also creates problems for the retailer comparing prices quoted by various wholesalers for their different selling grades if, as is increasingly the case, the retailer is personally unable to go and inspect individual batches of carcasses.

In Britain, carcass grading systems of this type have been developed by wholesalers buying from producers on a dead weight, rather than a live weight, basis. These have been used to encourage production of the kind of carcass they desire and discourage those which are not required. Such grading systems are particularly well developed for pigs where there has been a long history of dead weight selling by producers, but they are becoming used more commonly for cattle and sheep.

The diversity of demand in Britain reflected by, for example, the different carcass types embraced by the top grades operated by individual firms makes it difficult in the short- to medium-term to envisage acceptance of national grading schemes for cattle, sheep and pigs. This is in contrast to the discipline created by national grading schemes which have been operating in countries such as New Zealand and Denmark which rely on a strong export trade for much of their home production. It also contrasts with national grading schemes which have been operating in a number of countries primarily for internal trade. The best known of these schemes is probably that operated by the USDA for beef, some details of which will now be discussed briefly.

The USDA scheme is, in fact, in two parts: determination of the 'quality' grade and determination of the 'yield' grade. Carcasses submitted for quality grading must be well chilled and 'ribbed' (i.e. partially quartered to reveal the $m.l.d.$). The grader is required to assess the palatability indicating characteristics of the lean. To do this, he assesses the degree of marbling or intramuscular fat development from the $m.l.d.$ and maturity from the state of ossification of the cartilages, the colour of the bone and the colour and texture of the lean. These are then combined to form one of the following 8 grades according to a set method: prime, choice, good, standard, commercial, utility, cutter and canner.

The basis of the combination of the characteristics to form the premium grades is that as the animal becomes physiologically older there is a need for more marbling to compensate for the deleterious effects of increasing age on the eating characteristics of the meat. Until early 1976, when the scheme was modified, more marbling was required for carcasses to attain the top grades, but following evidence that high levels of marbling are not required for satisfactory eating quality, the marbling requirements were reduced. At the same time, the role of conformation in modifying the quality grade was dropped in view of mounting evidence about its irrelevance in this context.

The USDA yield grading system was introduced originally as a supplement to the quality grade, but it is now carried out whenever a carcase is 'quality' graded. The purpose of the yield grade is to separate carcasses into classes (5 in number) according to their expected yield of high-priced retail cuts after deboning and trimming to 9 mm of external fat. The estimation of the yield involves measuring the fat thickness over the $m.l.d.$ with an adjustment for unusual distribution, area of $m.l.d.$, carcass weight and the heart, kidney and pelvic fat percentage. In practice, the prediction is carried out subjectively by trained graders, although the standards are written in terms of the actual measurements.

An important way of overcoming the problems in Britain and other countries associated with individual traders wishing to operate their own private grading schemes is to develop and operate carcass classification schemes. Such schemes simply provide a method of describing the characteristics of carcasses which are of prime importance to the retailer cutting up and preparing meat for sale. The characteristics are described separately without attributing relative importance or cash value differences to them. Pig carcasses are described primarily in terms of their weight and fat thickness measurements, and beef and lamb carcasses by their weight, fatness and conformation (the latter two characteristics being assessed subjectively).

In view of the evidence presented earlier, it is relevant at this point to indicate the reasoning behind the inclusion of a judgement of conformation in these classification schemes. If one considers beef, the justification for including conformation is not only that it makes the beef scheme more readily acceptable to the meat trade by, for example, providing some indication of thickness of meat but that, amongst animals with similar percentages of SF, conformation appears to identify breeds with a better ratio of lean meat to bone and of SF to IF than other breeds. The Limousin with above average conformation in GB is an example of a breed at one extreme with a good lean to bone ratio and SF to IF ratio while the Canadian Holstein with below average conformation is an example of a breed at the other extreme. The Limousin also appears to have a somewhat higher percentage of lean meat in the more valuable cuts than breeds such as the Holstein. This role for conformation is of importance, since the breed of an animal is not normally known to the classifier of carcasses or in subsequent trading. Some further discussion of this subject is contained in MLC Technical Bulletin No. 18.

Given the framework of the descriptive information provided by classification, the buyer or seller can draw up his own grades by grouping the separate classification descriptions together and then attaching prices to these different groupings or grades. In a wider context, descriptive classification systems can be used as a basis for national grading schemes, such as the mandatory classification system for beef which will be operating in France from 1977 under the auspices of ONIBEV.

Classification schemes, then, provide a common link for describing important characteristics in a way which all can understand. They help producers to appreciate more clearly what the market requires, and the regular feedback of classification results to producers linked to wholesalers' buying schedules enables them to know more clearly the types of carcasses which they are producing and those in demand by retailers. The operation of such descriptive schemes should provide a sounder basis for livestock improvement in future.

V. Live Animal Evaluation

So far, the discussion has been about assessing the carcass characteristics of animals after slaughter, but in many situations it is important to obtain an estimate of these characteristics in the live animal. For example, it can be of value in selecting between and within breeds on farms or central performance testing stations, in selecting animals for slaughter on a fatness basis from breed comparison trials and progeny tests, or in the selection of commercial animals for slaughter.

It was noted earlier that weight on its own is not a sufficiently reliable predictor of carcass composition in view of the important variations in composition which exist amongst animals of similar weight. It follows that the value of alternative techniques for assessing composition in the live animal should be tested amongst animals of similar weight. Unfortunately, it is difficult to make meaningful comparisons between alternative procedures because many of the tests carried out in the past have not had the results adjusted to take out the effect of carcass weight variations.

The technique chosen for any particular purpose will depend on the requirements and constraints associated with the project. Unless the technique is to be applied to small numbers of animals under laboratory conditions it is important for it to be capable of being applied quickly and easily under field conditions and the equipment must be robust and transportable. In addition, it should not cause harm to the animal or depreciate the value of the resulting carcass.

The simplest technique is to handle the animal and assess its degree of fat development and conformation. It has already been observed that after allowing for fatness differences, judgement of conformation within a breed does not give a good guide to the characteristics which have been attributed to it over the years. So far as fatness estimation is concerned, there are undoubtedly some who are able to assess the degree of SF development quite accurately, but a good deal of training is required to reach the necessary expertise to enable such a technique to be used to distinguish amongst animals differing in fatness. The technique is particularly difficult where a range of breed types is included since, for example, some of the late-maturing European beef breeds give the impression of being fatter than they actually are. In many cases, it may be necessary to have the same standard operating in more than one place at the same time and over a period of several months or even years. Under these conditions, even with a highly skilled judge, there are likely to be advantages in having an objective assessment.

A wide range of objective techniques for assessing composition in the live animal have been developed over the years, and some of these have been reviewed by *Houseman* (1972). Some of the techniques available will be discussed briefly below.

A. Backfat Probes

These are used, primarily on pigs, to measure the thickness of SF at selected sites. A simple metal ruler with knife edge was the first probe used for this purpose (*Hazel and Kline*, 1952), but *Andrews and Whaley* (1955) developed the 'Leanmeter' probe which works on the difference in electrical conductivity between fat and lean. Good agreement has been reported between the two probing methods, but *Pearson et al.* (1957) considered that the simple metal ruler probe was the more reliable tool for estimating carcass leanness.

B. Ultrasonics

The value of estimating the fatness of pigs through the use of simple ultrasonic or echo-sounding equipment to measure fat depths is well accepted for pigs, where a large proportion of the total fat is subcutaneous. The equipment most commonly used for this purpose, at least in Europe, is of the A-mode type. Due to difficulties in identifying the ultrasonic signals relating to particular muscle boundaries, its value in estimating muscle depths and areas in pigs is not so good.

Various groups of workers in the USA, Germany and Denmark have attempted over the last decade or more to develop ultrasonic techniques to improve on the estimation of fatness, especially in beef where less of the total fat is subcutaneous than in pigs, and to improve the estimation of the underlying muscles. Some of the equipment developed works on the principle of linear scanning, such as the B-Scan equipment developed in the USA and sold under the trade name 'Scanogram'. Results of its use in beef have been reported by, for example, *Gillis et al.* (1973) in Canada and by *Tulloh et al.* (1973) in Australia. The former group carried out some comparative work with A-mode equipment and indicated the importance of having skilled operators. Some work using the Scanogram on lamb has been reported by *Kempster et al.* (1977).

The most sophisticated equipment evolved so far is probably that developed in Denmark in recent years. It permits some compound scanning and a description of the technique and results of its use have been reported on pigs and beef by *Andersen et al.* (1970), and *Andersen* (1975). In theory, such equipment should define SF and the underlying musculature more clearly than linear scanning equipment. More recent work in Denmark has led to the development of the 'Dan Scanner' which consists of an array of transducers on a fixed arc and which provides an almost instantaneous picture of the underlying tissues. While it does not provide a compound scan which will tend to limit the extent to which it can give a very good definition of the musculature, the speed with which it produces ultrasonic pictures could be advantageous in minimising the problem of body movement during scanning.

Ultrasonic equipment developed so far still leaves a good deal of scope for improving the definition of muscle boundaries and this is also likely to have an effect on the definition of fat. In the long run, it should be possible, by taking a series of scans along and across the back of an animal, to build up an overall picture of the composition of most of the carcass and, at the same time, provide information on the thickness of muscle and of the thickness and distribution of SF.

Measurements of the velocity of ultrasound through the limbs of living animals have been suggested as a method of estimating lean:fat ratios at suitable sites in the limbs (*Miles and Fursey*, 1974). Such a method has the advantage of objectivity whereas echo-sounding techniques of the types described above are subject to some errors of interpretation. Perhaps a combination of the two approaches might be valuable for increasing the precision of estimating fatness.

C. X-Rays

This method has been examined by, for example, *Dumont and Destandau* (1964) for its value in estimating fatness, but there are problems in taking satisfactory X-ray pictures of the living animal and of obtaining tissue measurements from the pictures. However, the technique could provide estimates of overall carcass composition and of the distribution and thickness of the tissues.

Recently, EMI Medical Limited have developed primarily for diagnostic work in humans a body scanner using a non-conventional X-ray system. It has been heralded as the greatest diagnostic advance since the discovery of X-rays. It could prove to be a valuable tool for assessing the composition of the animal body, but with its present very high cost and the need to bring animals to a central point for measurement, the scope for using it on animals seems limited in the near future.

D. Tracer Dilution Techniques

The techniques involve administering a known amount of tracer which will become uniformly distributed throughout a compartment in the animal body. The concentration of the tracer at equilibrium is then measured.

Several of the dilution techniques are used to estimate the amount of body water which is related to total fat content. Tritiated water, deuterium oxide, antipyrene and related 4-amino-antipyrene and *N*-acetyl-4-amino-antipyrene are among the tracers which have been used to estimate body water in beef, sheep and pigs. These techniques have given conflicting results in terms of their ability to estimate variations in carcass fatness.

As already indicated, a crucial test of the techniques is their ability to distinguish fatness amongst carcasses of similar weight. In this situation, a number of techniques may not prove valuable (*Cuthbertson et al.*, 1973) although they may be better associated with the chemical fat content of the body than with the physically separable fat. One of the problems of estimating body water in the carcass from measurements on the live animal, particularly in the case of ruminants, is the effect of variation in the water content of the alimentary tract. However, even if body water estimations could give satisfactory estimates of total carcass fat, they cannot provide information on the distribution and thickness of the tissues in different parts of the carcass. Similar deficiencies in estimating carcass variations are associated with the other dilution techniques referred to below.

One of these techniques involves measuring the uptake of a tracer substance which dissolves uniformly in fat. The animal is enclosed in a chamber and a known amount of gas introduced. When equilibrium is reached, the amount of gas absorbed can be calculated. Several gases have been tried, including krypton (*Hytten et al.,* 1966), but there have been problems associated with their application.

Several investigations have indicated a relationship between body composition and various volume measurements of blood and its components. A number of techniques have been tried, including the estimation of red cell volume by ^{51}Cr dilution (*Doornenbaal et al.,* 1962). This technique is based on the relationship which exists between the oxygen-transporting vehicle and the oxygen-consuming tissues, largely the lean meat mass.

Another technique involving dilution relies on the fact that much of the potassium in the body lies in the lean meat mass, so that estimation of the former in the body provides a measure of the latter. *Houseman* (1972) reported the use of ^{42}K as a tracer for this purpose.

E. Density

This approach involves estimating body volume and a number of methods have been tried. These have included air displacement and helium dilution, but there have been problems in, for example, measuring changes in volume, particularly the head space, with sufficient precision. A different approach has been to determine the volume of the animal from contour lines produced by photogrammetry equipment, but this technique is expensive. A cheaper, but not, perhaps, as accurate a procedure for obtaining the contour lines of the body is the use of the moiré method described by *Speight et al.* (1974). Briefly, the method involves the casting of a shadow of a plane, equispaced line grating onto the object by means of a point light source. The moiré pattern produced is a system of contour lines of equal depth from the grating provided by the interaction of two-line systems (shadow and grating). Once the contour lines have been photographed, there still remains the task of deriving volume measurements. As with dilution and other density techniques, there are difficulties in drawing conclusions on carcass composition because of the effect of non-carcass parts, and they cannot provide information on the thickness and distribution of the tissues.

F. Potassium-40

The use of potassium to assess leanness has already been noted. A non-dilution method which is used involves measuring the naturally occurring isotope, ^{40}K. The animal is shielded, as far as possible, from any external radiation and the ^{40}K radiation from the animal is measured. The technique has been tried extensively in the USA (*Frahm et al.,* 1971). However, in spite of its use, for example, in Oklahoma for the evaluation of pig and beef breeding stock, doubts about its value remain in relation to the high cost of the equipment.

G. Electromagnetism

In recent years, equipment has been developed in the USA and sold under the trade name 'EMME' which estimates body composition by the difference in electrical conductivity between lean and fat. *Domermuth et al.* (1973), and *Koch and Varnadore* (1976) are among those who have reported some results of its use, but further evaluation is necessary. The technique is quick to operate and is non-destructive.

H. Lean Tissue Growth Rate

Stimulated by *Kielanowski* (1966), an increasing amount of interest has been shown in the use of growth performance characteristics, such as daily live weight gain or food conversion ratio, to predict the lean content of live animals particularly in relation to boar performance testing. It has been suggested that where pigs are performance tested in such a way that they are fed a fixed amount of feed each day and slaughtered at a fixed age, their live weight at the end of test will be a direct reflection of their genetic potential for lean meat gain and also the percentage of lean in the carcass. The argument goes on that since maintenance requirements are closely related to days on test, it will be approximately the same for all pigs grown for the same period of time. If feed intake per day is also the same for all pigs, the energy available for growth will also be approximately the same for all pigs. Then, since it requires some three times as much energy to deposit fat as it does to lay down lean meat, the animals which are heaviest at the end of the test will be those which lay down lean meat at the fastest rate.

Such a system, if proven, provides a simple method of testing without the necessity of measuring carcass characteristics related to leanness. However, there are several disadvantages to set against this potential advantage. Firstly, it may still be necessary to carry out carcass evaluation of progeny or sibs to measure other characteristics such as muscle thickness and quality. Secondly, the test will not be representative of commercial practice where animals are normally slaughtered at a fixed live weight. Thirdly, since such a system of testing requires restricted feeding, it may not give the best results in terms of the genetic improvement made.

I. Physiological Indices

Research work is now in progress at a number of centres to provide a better understanding of the physiological control of the development of lean meat and of fat. Such work may lead to the identification of some physiological measurements which might be used to predict carcass merit. *Gregory et al.* (1976) have reported favourably on a study of insulin metabolism and *Bakke* (1975) on the use of measures of fat metabolism. Studies of this type, relating to the physiological control of body development, seem likely to become of more importance in future.

A good deal of work has been carried out both in ruminants and non-ruminants on some of the foregoing techniques for evaluating the carcass composition of live animals. However, very little work has been done to compare the value of the alternative techniques on the same sample of animals. One of the few attempts to test out several techniques on the same sample has been carried out on pigs by *Houseman* (1972), but the study was limited to only 24 pigs varying widely in fatness. There is, therefore, a need for further work of this nature.

In any comparative study of techniques, it is important to examine their value, individually and in different combinations, over production characteristics. The techniques included in any comparative study will depend on the circumstances in which the techniques will be used. For example, in the UK it appears that legislation would prevent the use of certain of these techniques (e.g. backfat probes and injection of tracers) for uses other than for diagnostic purposes. Even if they were permitted for routine use in, for example, performance-testing animals, it is doubtful whether some of them would be feasible for routine use under field conditions. Taking into account the foregoing restrictions, then the following techniques could be worth comparing in the short term at least under UK conditions: subjective assessment, ultrasonics, density, X-rays, ^{40}K, lean tissue growth rate and, perhaps, electromagnetism. Of these, the ultrasonic scanning technique coupled with growth performance data seems most likely in the current state of knowledge to provide the information required. Apart from the ultrasonic and X-ray techniques, the others mentioned provide no information on the distribution and thickness of the tissues which could be of importance in the future as intensive selection is applied to specific characteristics.

VI. Conclusion

This chapter has tried to give the reader an insight into the characteristics of carcasses which are of importance and how variations in these may be assessed under different conditions. With economic forces increasing the pressure to improve the efficiency of meat production, it will become more important to choose the best possible evaluation procedure for each set of circumstances.

The evidence presented in the chapter goes some way to indicating the current state of knowledge about the procedures most appropriate in, for example, studying the carcass characteristics of different breeds or of describing commercial carcasses under fast-moving slaughter-line conditions. In order to have more confidence in applying particular evaluation techniques in different circumstances, there is a need for more information on the extent to which each technique is reliable when applied over a wide range of carcass types.

There is much merit in being able to make accurate estimates of body composition in the live animal. It would be of value, for example, to be able to study changes in composition as the animal grows and to examine how these changes are influenced by genetic or nutritional effects. While there is a wide range of alternative procedures for assessing composition in the live animal, there is a dearth of information on their relative merits. To remedy this, there appears to be a need for some of the most likely alternatives to be tested on the same sample of animals.

There is growing interest in the physiological control of body composition. The results of hormone assays are now being related to body composition and, in the long-term, since growth and development are very dependent on hormonal control, it may be possible to manipulate hormone levels to achieve the desired carcass characteristics. This area of study appears to be one which should command a good deal of research activity in the years ahead.

References

Adam, J.L. and Smith, W.C.: The use of specific gravity and its reciprocal in predicting the carcass composition of pigs slaughtered at three weights. Anim. Prod. *6:* 97–105 (1964).

Andersen, B.B.: Recent experimental development in ultrasonic measurement of cattle. Livest. Prod. Sci. *2:* 137–146 (1975).

Andersen, B.; Pedersen, O.K.; Busk, H.; Lund, S.A. und Jensen, P.: Ultraschallmessungen bei Rindern und Schweinen. Fleischwirtschaft *50:* 843–846 (1970).

Andrews, F.N. and Whaley, R.M.: Measure fat and muscle in live animal and carcass. Rep. Ind. agric. exp. Sta. *68:* 27–29 (1955).

AOAC: Official methods of analysis of the Association of Official Analytical Chemists; 12th ed. (Association of Official Analytical Chemists, Washington 1975).

Bakke, H.: Serum levels of non-esterified fatty acids and glucose in lines of pigs selected for rate of gain and thickness of back fat. Acta agric. scand. *25:* 113–116 (1975).

Berg, R.T. and Butterfield, R.M.: New concepts of cattle growth (Sydney University Press, Sydney 1976).

Boccard, R. et Radomska, M.J.: Etude de la production de la viande chez les ovins. VI. Influence de la forme du membre postérieur sur ses caractéristiques technologiques. Annls Zootech. *12:* 5–15 (1963).

Brayshaw, G.H.; Carpenter, E.M., and Phillips, R.A.: Butchers and their customers. Report No. 1, Department of Agric. Marketing (University of Newcastle-upon-Tyne, Newcastle-upon-Tyne 1965).

Butterfield, R.M.: Relative growth of the musculature of the ox. Symp. Carcass Composition and Appraisal of Meat Animals, CSIRO, Melbourne 1963.

Carroll, M.A. and Conniffe, D.: Beef carcass evaluation: fat, lean and bone. Growth and development of mammals, Proc. 14th Easter School in Agricultural Science, University of Nottingham 1967, pp. 389–399 (Butterworths, London 1968).

Commission of the European Communities: Criteria and methods for assessment of carcass and meat characteristics in beef production experiments (Commission of the European Communities Directorate General, Luxembourg 1976).

Cook, G.L.; Cuthbertson, A.; Smith, R.J., and Kempster, A.J.: Prediction of pig carcass composition by sample joint dissection and fat thickness measurements. Proc. Br. Soc. Anim. Prod. *3:* 86 (1974).

Cuthbertson, A.; Harrington, G., and Smith, R.J.: Tissue separation – to assess beef and lamb variation. Proc. Br. Soc. Anim. Prod. *1:* 113–122 (1972).

Cuthbertson, A.; Read, J.L.; Davies, D.A.R., and Owen, J.B.: Performance testing ram lambs. Proc. Br. Soc. Anim. Prod. *2:* 83 (1973).

Domermuth, W.F.; Veum, T.L.; Alexander, M.A.; Hedrick, H.B., and Clark, J.L.: Evaluation of EMME for swine. J. Anim. Sci. *37:* 259 (1973).

Doornenbaal, H.; Asdell, S.A., and Wellington, G.H.: Chromium-51-determined red cell volume as an index of 'lean body mass' in pigs. J. Anim. Sci. *21:* 461–463 (1962).

Dumont, B.L. et Destandau, S.: Comparaison de quatre méthodes de mesure de l'épaisseur des tissus adipeux sous-cutanés chez le porc vivant. Annls Zootech. *13:* 213–216 (1964).

EAAP: Beef carcasses – methods of dressing, measuring, jointing and tissue separation. Publication No. 18, European Association of Animal Production (Meat Research Institute, Bristol 1976).

Florence, E. and Mitchell, K.G.: A procedure for preparation of pig carcasses for chemical analysis with special reference to micro-analysis. Proc. Br. Soc. Anim. Prod. *1:* 101–107 (1972).

Frahm, R.R.; Walters, L.E., and McLellan, C.R.: Evaluation of ^{40}K count as a predictor of muscle in yearling beef bulls. J. Anim. Sci. *32:* 463–469 (1971).

Gillis, W.A.; Burgess, T.D.; Usborne, W.R.; Greiger, H., and Talbot, S.: A comparison of two ultrasonic techniques for the measurement of fat thickness and rib eye area in cattle. Can. J. Anim. Sci. *53:* 13–19 (1973).

Gregory, N.G.; Wood, J.D., and Lister, D.: Studies of the role of plasma insulin in controlling body composition and meat quality in pigs. Anim. Prod. *22:* 150 (1976).

Harrington, G.: The relative accuracy and cost of alternative methods of pig carcass evaluation. Z. Tierzucht. ZuchtBiol. *82:* 187–198 (1965).

Harrington, G. and Kempster, A.J.: An examination of the fatness of commercial British beef carcasses and estimates of waste fat production. Anim. Prod. *24:* 152 (1977).

Hazel, L.N. and Kline, E.A.: Mechanical measurement of fatness and carcass value on live hogs. J. Anim. Sci. *11:* 313–318 (1952).

Houseman, R.A.: Studies of methods of estimating body composition in the living pig; PhD thesis, Edinburgh (1972).

Hytten, F.E.; Taylor, K., and Taggart, N.: Measurement of total body fat in man by absorption of ^{85}Kr. Clin. Sci. *31:* 111–119 (1966).

Kempster, A.J.; Avis, P.R.D.; Cuthbertson, A., and Harrington, G.: Prediction of the lean content of lamb carcasses of different breed types. J. agric. Sci. *86:* 23–24 (1976a).

Kempster, A.J.; Avis, P.R.D., and Smith, R.J.: Fat distribution in steer carcasses of different breeds and crosses. 2. Distribution between joints. Anim. Prod. *23:* 223–232 (1976b).

Kempster, A.J.; Cuthbertson, A., and Harrington, G.: Fat distribution in steer carcasses of different breeds and crosses. 1. Distribution between depots. Anim. Prod. *23:* 25–34 (1976c).

Kempster, A.J.; Cuthbertson, A.; Jones, D.W., and Owen, M.G.: A preliminary evaluation of the Scanogram for predicting the carcass composition of live lambs. Anim. Prod. *24:* 145 (1977).

Kempster, A.J.; Cuthbertson, A., and Smith, R.J.: Variation in lean distribution among steer carcasses of different breeds and crosses. J. agric. Sci. *87:* 533–542 (1976d).

Kempster, A.J. and Jones, D.W.: Relationships between the lean content of joints and overall lean content in steer carcasses of different breeds and crosses. J. agric. Sci. *88:* 193–201 (1977).

Kielanowski, J.: Conversion of energy and the chemical composition of gain in bacon pigs. Anim. Prod. *8:* 121–128 (1966).

Koch, R.M. and Varnadore, W.L.: Use of electronic meat measuring equipment to measure cutout yield of beef carcasses. J. Anim. Sci. *43:* 108–113 (1976).

McClelland, T.H. and Russel, A.J.F.: The distribution of body fat in Scottish Blackface and Finnish Landrace lambs. Anim. Prod. *15:* 301–306 (1972).

Miles, C.A. and Fursey, G.A.J.: A note on the velocity of ultrasound in living tissue. Anim. Prod. *18:* 93–96 (1974).

Osinska, Z.: Problems in the evaluation of pig carcasses. Growth and development of mammals. Proc. 14th Easter School in Agricultural Science, Nottingham 1967, pp. 416–426 (Butterworths, London 1968).

Pearson, A.M.; Price, J.F.; Hoefer, J.A.; Bratzler, L.J., and Magee, W.T.: A comparison of the live probe and lean meter for predicting various carcass measurements of swine. J. Anim. Sci. *16:* 481–483 (1957).

Pomeroy, R.W.; Williams, D.R.; Harries, J.M., and Ryan, P.O.: Composition of beef carcasses. 1. Material, measurements, jointing and tissue separation. J. agric. Sci. *83:* 67–77 (1974).

Preston, T.R. and Willis, M.B.: Intensive beef production; 2nd ed. (Pergamon Press, Oxford 1974).

Rhodes, D.N.: Eating quality of meat – the interaction of composition, preference, regulation and marketing. Proc. 22nd Eur. Meet. Meat Research Workers, Malmö 1976.

Richmond, R.J. and Berg, R.T.: Muscle growth and distribution in swine as influenced by liveweight, breed, sex and ration. Can. J. Anim. Sci. *51:* 41–49 (1971).

Speight, B.S.; Miles, C.A., and Moledina, K.: Recording carcass shape by a moiré method. Med. Biol. Engng *8:* 221–226 (1974).

Timon, V.M.: Genetic studies of growth and carcass composition in sheep. Growth and development of mammals. Proc. 14th Easter School in Agricultural Science, Nottingham 1967, pp. 400–415 (Butterworths, London 1968).

Timon, V.M. and Bichard, M.: Quantitative estimates of lamb carcass composition. 1. Sample joints. Anim. Prod. *7:* 173–181 (1965a).

Timon, V.M. and Bichard, M.: Quantitative estimates of lamb carcass composition. 2. Specific gravity determination. Anim. Prod. *7:* 183–187 (1965b).

Tulloh, N.M.; Truscott, T.G., and Lang, C.P.: An evaluation of the Scanogram for predicting the carcass composition of live cattle. Report submitted to the Australian Meat Board (School of Agriculture and Forestry, Melbourne 1973).

Alastair Cuthbertson, Meat and Livestock Commission, PO Box 44, Queensway House, Queensway, *Bletchley, Milton Keynes MK2 2EF* (England)

Subject Index

World Review of Nutrition and Dietetics

Contents of Volumes 5–27

Contributor Index for Volumes 5–27

The numbers following the names refer to the volumes